Questions and Answers on
Electric Wiring

Related titles in the Questions and Answers series

Questions and Answers on

Electric Wiring

HENRY A. MILLER
C.G.I.A., C.Eng., M.I.E.E., F.R.S.A.

LONDON
NEWNES—BUTTERWORTHS

THE BUTTERWORTH GROUP

ENGLAND
Butterworth & Co (Publishers) Ltd
London: 88 Kingsway, WC2B 6AB

AUSTRALIA
Butterworth Pty Ltd
Sydney: 586 Pacific Highway, NSW 2067
Melbourne: 343 Little Collins Street, 3000
Brisbane: 240 Queen Street, 4000

CANADA
Butterworth & Co (Canada) Ltd
Scarborough: 2265 Midland Avenue,
Ontario M1P 4S1

NEW ZEALAND
Butterworths of New Zealand Ltd
Wellington: 26–28 Waring Taylor Street, 1

SOUTH AFRICA
Butterworth & Co (South Africa) (Pty) Ltd
Durban: 152–154 Gale Street

First published in 1974 by
Newnes-Butterworths, an imprint of
The Butterworth Group

ISBN 0 408 00152 6

Filmset by Keyspools Ltd, Golborne, Lancashire

*Printed in Great Britain by Butler & Tanner Ltd,
Frome, Somerset.*

PREFACE

The term *electric wiring* in its broadest sense embraces the installation of electrical equipment in buildings generally, including
control and protection. This book is designed to give relevant
information concerning wiring simply and in logical sequence in a
question and answer form which facilitates reference.

Since electrical installation work in the U.K. should be carried
out in accordance with the latest edition of the *Regulations for the
Electrical Equipment of Buildings* issued by the Institution of
Electrical Engineers, the requirements of the Regulations are
identified with current practice, and references to them must
constitute an important part of any book on wiring.

Acknowledgement is made to the Institution of Electrical
Engineers for permission to make use of the Regulations in this
respect. Also to the companies named beneath certain of the
illustrations, who have supplied information concerning their
products.

Henry A. Miller

CONTENTS

1

INTRODUCTION

What is electric wiring?

Electric wiring is a means by which a consumer of electric energy can operate his items of electrical equipment as and when he wishes. The wiring extends from the consumer's terminals, where it is connected to the electricity supply mains, to the various outlet points such as ceiling roses and wall sockets. The term *fixed wiring* is often used to distinguish the wiring serving the fixed outlet points from flexible wiring such as that attached to appliances, familiarly known as leads.

Wiring comprises *cables* together with associated apparatus for control and protection. Cables consist of conductors, insulation and sometimes mechanical protection. The purpose of a conductor is to carry the current, and it is therefore constructed of a material offering little resistance to the flow of electric current, such as copper or aluminium. The conductor is generally in the form of either a single wire or of a group of wires (known as *strands*) in contact with each other.

The function of insulation is to prevent the current from leaking away from the conductor at places where it is not required to flow. Thus, a material which offers an extremely high resistance to the flow of current is used for the insulation, such as polyvinyl chloride (p.v.c.) or magnesia (mineral).

Mechanical protection may take the form of an integral overall sheath of insulating material or metal, or the cable may be enclosed within a conduit, trunking or duct. Two or more insulated conductors included within a cable and provided with mechanical protection are each described as a *core*.

9

The form of protection used largely determines what is described as the *system of wiring*. Typical systems of wiring include p.v.c.-insulated p.v.c.-sheathed cables, insulated cables within conduits (either metallic or nonmetallic), and mineral-insulated metal-sheathed cables.

Systems of wiring should not be confused with *methods of wiring*, such as the looping-in method employed when wiring is within conduits, and the joint box method applicable to sheathed wiring.

What are the Wiring Regulations?

The I.E.E. Wiring Regulations, or more correctly the *Regulations for the Electrical Equipment of Buildings*, are principally concerned with the safety of electric wiring and electrical installations generally. They are recognised as a U.K. national code in this respect. The Wiring Regulations Committee of the Institution of Electrical Engineers frames the Regulations and amends them from time to time as necessary in the light of new developments. The Institution also publishes a *Guide to the I.E.E. Wiring Regulations* which aims to point out some of the reasoning behind the Regulations and their implications in day-to-day terms.

The Regulations relate mainly to requirements for installation, testing and maintenance of consumers' wiring and equipment, but certain requirements for the construction of electrical equipment are included, mainly in the form of references to British Standards. For guidance on good practice in certain aspects of electrical installations beyond the scope of the I.E.E. Regulations, reference can be made to the British Standard Codes of Practice issued by the British Standards Institution.

What is the standard U.K. electricity supply?

Except for large consumers, the standard electricity supply in this country is 3-phase 4-wire alternating current at a frequency of 50 hertz, 415/240 volts r.m.s., with earthed neutral. This means that there is a pressure of 415 volts between any two phases, and 240 volts between any one phase and the neutral. Successive phases are designated red, yellow and blue. The arrangement is shown in

Fig. 1, the star point of the substation transformer secondary winding being earthed and forming the neutral.

Ordinary domestic supplies and those for small consumers usually comprise one phase and neutral, and are described as 240 volt, single-phase 2-wire, so that the service cable by which the supply is

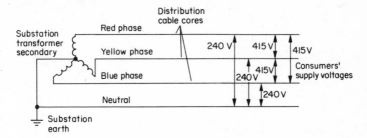

Fig. 1. Standard supply voltages

delivered to the premises is 2-core. Supplies to larger buildings, such as workshops, supermarkets, schools, etc., are generally 415/240 volt 3-phase 4-wire, which means that the service cable has four cores, and the single-phase loads such as lighting and heating are balanced over the three phases. Very large consumers, such as large factories and hospitals, are often provided with substation transformers within their own confines.

How is the supply system arranged?

Electricity supply to the public can be conveniently considered in three stages: generation, transmission and distribution (Fig. 2). Generation takes place in power stations which contain alternators driven by turbines. The generated electrical energy is transformed to a very high voltage (264 or 400 kilovolts) and transmitted across country by overhead lines. At towns and other highly populated areas, the transmitted energy is taken to substations and transformed to the standard supply voltage. From the substations, the energy is distributed by underground cables along the roads, and from there it is taken by service cables to consumers' premises.

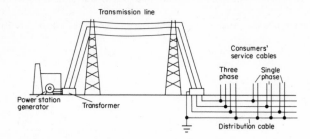

Fig. 2. Arrangement of electricity supply

What is the intake position?

Sometimes called the 'service' position, this is the point in a building at which the supply undertaking (e.g. the area electricity board) terminates the service cable. It is usually the situation of the main controlling and protective equipment which may vary from a simple consumer's control unit in domestic premises to a large switchboard panel in an industrial estate. It usually includes the supply undertaking's meters.

What is the service head?

The service head comprises the termination of the service cable. In the case of a paper-insulated lead-sheathed and armoured service cable, it includes the gland, armour clamp, sealing chamber, service fuse and neutral link.

What control and protective equipment is required?

The I.E.E. Wiring Regulations require every consumer's installation to be adequately controlled by switchgear readily accessible to the consumer. The equipment must incorporate: (*a*) means of isolation, (*b*) means of excess-current protection, and (*c*) means of earth-leakage protection.

12

Means of isolation may consist of a linked switch or linked circuit-breaker capable of operating on load. The means of excess-current protection may be either a fuse, which is a device for opening a circuit by means of a conductor designed to melt when an excessive current flows, or a circuit-breaker which is a mechanical device for making and breaking a circuit both under normal conditions and under abnormal conditions, such as those of a short-circuit, the circuit being broken automatically.

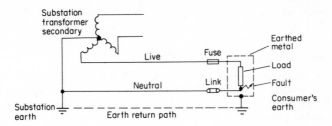

Fig. 3. Path of leakage current through fault, earthed metal, consumer's earth, earth return path, earthed neutral of transformer, transformer secondary, etc.

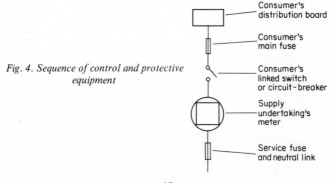

Fig. 4. Sequence of control and protective equipment

13

While fuses or circuit-breakers can always be used for protection against overload or short-circuit, they may only be used for earth-leakage protection where the impedance of the leakage current path (see Fig. 3) is sufficiently low to conduct enough current to operate the fuse or circuit-breaker. This is known as direct earthing. Where it is impossible to obtain a sufficiently low impedance, it is necessary to install an earth-leakage circuit-breaker.

The arrangement of the equipment controlling the supply to an installation must be such that the means of isolation follows the consumer's terminals directly. In most cases, the linked switch or circuit-breaker will be connected to the outgoing terminals of the supply undertaking's metering equipment, as shown in Fig. 4.

What arrangements are made for metering?

While the controlling and protective equipment at the intake is being wired up by the person carrying out the electrical installation work, provision has to be made for connection of the supply under-taking's meters. This provision takes the form of cables, known as meter tails, loops or bights, extending between the outgoing terminals of the service head and the consumer's main control.

The number of meter tails to be provided depends on the nature of the installation. For a domestic consumer intending to pay in accordance with a two-part tariff (i.e. a fixed quarterly charge plus so much per unit irrespective of whether the electricity is used for lighting, heating or cooking, etc.), only one pair of tails would need to be left. For a commercial or industrial consumer paying one price per unit for lighting and another price per unit for heating, two pairs of tails would be required, with the lighting and heating installations separately controlled. Separate provision would have to be made for a load supplied only at off-peak periods (e.g. storage heating).

What is earthing?

The term *earthing* means effective connection to the general mass of the earth. The primary requirement is an *earth electrode* which is in contact with earth. The earth electrode may take the form of a metal rod or plate, a system of underground metal pipes, or other means of effective earth connection.

The final conductor at one end of which connection is made to the earth electrode is called the *earthing lead* (Fig. 5). This terminates at its other end at the *consumer's earthing terminal* which is provided adjacent to the consumer's terminals.

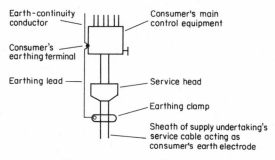

Fig. 5. Earthing terms illustrated

A conductor (including any clamp) which connects the earthed parts of an installation to the earthing terminal is called an *earth-continuity conductor*. It may consist wholly or partly of the metal conduit or duct, of the metal sheaths of the cables forming the wiring, or of a special earth-continuity conductor in a cable.

The following must be connected to an earth-continuity conductor: (*a*) exposed metalwork of all apparatus (subject to certain exemptions), (*b*) the earthing terminal of every socket-outlet and spur box, (*c*) the earthing terminal of every lighting point and switch (unless in the form of an earthed metal box with the switch-plate fixed in electrical contact with it).

What is bonding?

In general, bonding means electrical connection, not normally for the purpose of carrying current, but so as to ensure a common potential. The idea is that if all metal within reach is at the same electrical potential, a person touching any part will not suffer a shock. The I.E.E. Regulations require that bonding must be carried out *after* connection of the consumer's earthing terminal to the means of earthing.

Bonding connection to gas or water services should be as close as possible to the point of entry of the services into the premises. Other

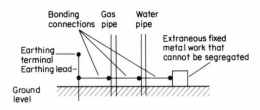

Fig. 6. Bonding

exposed metalwork which should be bonded includes baths, exposed metal pipes of all kinds, radiators, sinks, tanks and accessible structural steelwork (Fig. 6).

What is all-insulated construction?

This implies the construction of equipment in such a way that live conductors and parts that may become live under fault conditions cannot be touched. In the case of apparatus, it includes approved methods of guarding. It is regarded as an acceptable method of protecting apparatus against earth-leakage currents.

What is double insulation?

Double insulation generally applies to appliances and lighting fittings, and the specifications are given in the appropriate British Standards. The basic requirement is that there must be a covering of insulating material in addition to the insulation normally

16

provided, so that a breakdown of the normal insulation alone does not present any risk of shock.

What is protective multiple earthing?

Generally known as p.m.e., this is a method of connection with earth in which the earthing lead is connected to the consumer's earthing terminal which, in turn, is connected to the neutral conductor of the supply. The method is used where local earthing conditions are so poor that other forms of earth-leakage protection would be unsatisfactory.

Such a system involves all protected metalwork being connected by earth-continuity conductors to the neutral service conductor at the supply intake. Thus, any line-to-earth fault that may develop is also a line-to-neutral fault. Provided that the fault is not of very high resistance, the fault current would be sufficient to blow the fuse or cause operation of the excess-current circuit-breaker protecting the circuit concerned.

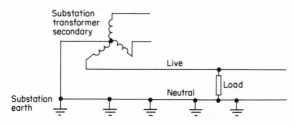

Fig. 7. Arrangement of protective multiple earthing

As shown in Fig. 7, the neutral conductor is earthed at a number of points on the system. There must be no single-pole switch, circuit-breaker or other control device, or removable link, in the neutral conductor. A break in the neutral conductor could have serious consequences. There must be no possibility of the neutral conductor attaining an appreciable potential with respect to earth, as this would extend the shock risk to all protected metalwork on all installations connected to the system.

Which items of equipment should be labelled?

A permanent label indelibly marked in lettering at least 44·75 mm high reading *Safety Electrical Earth—Do Not Remove* must be permanently fixed at every connection of an earthing lead to an earth electrode or other means of earthing.

Also, every switch or circuit-breaker, the purpose of which is not obvious, must be labelled to indicate the apparatus it controls.

Each distribution board must have an adjacent list indicating the circuit protected by each fuse or circuit-breaker and the appropriate current rating.

Every item of electrical apparatus or enclosure within which medium voltage (250–650 volts) exists must have notices at points of access warning of the maximum voltage present.

What is diversity?

In most cases it is extremely unlikely that an electrical installation would be fully loaded at any given time. Domestic consumers, for instance, do not usually use lighting, socket-outlets, electric oven, hotplates, kettle, water heating etc. all at once. It would under these circumstances be unreasonable to have to cater for mains equipment suitable for a total load which would never be achieved in practice. Therefore, the I.E.E. Regulations permit, where it can be justified, the application of a *diversity factor* which results in a reduction in the sizes of conductors and switchgear of circuits other than final subcircuits. An Appendix in the Regulations provides a guide to the amount of reduction where applicable.

Appropriate allowance for diversity is made in the Regulations in respect of final subcircuits for 13 A socket-outlets and cooking appliances.

How are cable sizes assessed?

In the case of final subcircuits it is rarely necessary to calculate the size of cable conductor required. For 5 A lighting final subcircuits we know that a 1 mm² (1/1·13) p.v.c.-insulated cable will usually suffice. For a 30 A ring final subcircuit, the minimum size is 2·5 mm²

18

(1/1·78). But for cables feeding distribution boards it is often necessary for the minimum size to be assessed using the tables of current rating and voltage drop included in the I.E.E. Regulations.

A cable conductor should be capable of carrying the continuous load current not only without undue overheating, taking the class of excess-current protection provided into account, but also without resulting in excessive voltage drop. The maximum permissible voltage drop is 2·5 per cent of the declared supply voltage which, in the case of a 240 volt supply, is 6 volts.

For example, supposing it is necessary to run a 35 mm twin p.v.c.-insulated and sheathed cable, nonarmoured, carrying 20 A and provided with coarse excess-current protection. From Table 3M of the Regulations, column 3, we see that the smallest size that could be used is 7/0·85, rated at 24 A. From column 4 of the same table we find that the voltage drop per ampere per metre for this cable is 10 mV (0·01 V). Thus, the voltage drop will be

$$0·01 \times 20 \times 35 = 7 \text{ volts}$$

This is higher than the permitted maximum of 6 V, and cannot therefore be used. The next larger size is 7/1·04, rated at 30 A, with a voltage drop per ampere per metre of 6·8 mV (0·0068 V). The voltage drop using this will be

$$0·0068 \times 20 \times 35 = 4·76 \text{ volts}$$

This is below the permitted 6 V, and this size is therefore suitable.

2

PROTECTIVE REQUIREMENTS

What is the excess-current protection?

When an electric current flows through a conductor, heat is generated. The quantity of heat produced depends on the value of the current and on the time for which it flows. The rate of heat production in relation to the rate of heat dissipation must never be such that an undue temperature rise is permitted, as this could damage the insulation. Therefore, if, owing to an overload or a fault, excessive current flows, there must be some form of protection to disconnect the circuit concerned before any damage can be caused.

Two recognised methods of excess-current protection consist of the insertion in each live conductor of (*a*) a fuse, or (*b*) a circuit-breaker with excess-current release.

What is close excess-current protection?

This is defined as excess-current protection which will operate within 4 hours at 1·5 times the designed load current of the circuit which it protects. Devices recognised by the I.E.E. Regulations as affording close protection include:

(*a*) BS 88 fuses fitted with fuse links marked to indicate a Class P or Class Q1 fusing factor;

(*b*) Fuses fitted with fuse links complying with BS 1361;

(*c*) Miniature and moulded case circuit-breakers complying with BS 3871;

(*d*) Circuit-breakers set to operate at an overload not exceeding 1·5 times the designed load current of the circuit.

What is coarse excess-current protection?

This is excess-current protection which will not operate within 4 hours at 1·5 times the designed load current of the circuit which it protects.

Devices recognised by the I.E.E. Regulations as affording only coarse current protection include:

(*a*) BS 88 fuses fitted with fuse links marked to indicate a Class Q2 or Class R fusing factor;

(*b*) Semi-enclosed (rewirable) fuses complying with BS 3036.

What is a fuse?

A fuse normally consists of a *fuse base* and a *fuse link*. The fuse base generally contains sockets designed to engage with contact pieces of the link. The fuse link usually takes the form of a carrier supporting the *fuse element*, which is a conductor intended to melt when excessive current flows and thus to isolate from the supply an overloaded or faulty circuit giving rise to the flow of excess current.

The fuse element may consist of wire connected to terminals in the carrier or, alternatively, of a totally enclosed manufactured element designed to be clipped into, or bolted to, the carrier.

The maximum current that a fuse will carry without undue deterioration is called the *current rating* of the fuse. The minimum current required to melt, or blow, the fuse is called the *fusing current*. The ratio of fusing current to current rating is known as the *fusing factor*.

Fuses may be incorporated in a distribution board to protect a given number of circuits. When employed on a single-phase supply, each fuse represents a *way*. An outgoing circuit connected to such a distribution board and feeding outlet points is known as a *final subcircuit*.

What is a semienclosed fuse?

A semienclosed fuse, shown in Fig. 8, is one in which the fuse element is a piece of wire of suitable cross section, which will conduct the rated current without overheating, but which will melt when excess-current of 1·5 to 2 times the rated current flows (i.e. it has a fusing factor of 1·5 to 2).

One of the disadvantages of a semienclosed fuse is that the surface of the element is subject to oxidisation, which affects its

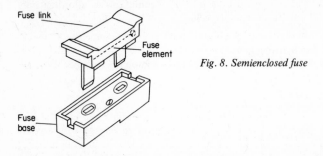

Fig. 8. Semienclosed fuse

characteristics. Another is that under heavy fault conditions there is a risk of flashover and consequent fire risk. Yet another is that it is not suitable for heavy currents or high voltages. Its main advantage is its cheapness.

What is a cartridge fuse?

As the name implies, this type has a fuse link in the form of a cartridge. The cartridge (see Fig. 9) contains an element which is

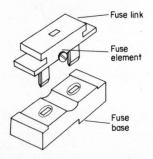

Fig. 9. Cartridge fuse

completely sealed. The element may consist of a bunch of wires or of a thin strip. Some of the characteristics of the element may be designed during manufacture (e.g. by cutting portions away or by inclusion of low melting-point inserts). Generally, the element is surrounded by fireproof material such as quartz, the whole being enclosed within a ceramic tube.

A cartridge fuse of the type known as *high-breaking capacity* can have a fusing factor as low as 1·1. Being sealed, it does not oxidise and does not present a fire risk when operating under heavy fault conditions. It is rapid in operation and can be designed for very heavy currents.

What is an excess-current circuit-breaker?

A circuit-breaker is a mechanical device for making and breaking a circuit both under normal conditions and under abnormal conditions such as those of a short-circuit, the circuit being broken automatically. When used for excess-current protection, it has a release mechanism operated by a trip coil, the operating current being prearranged (see Fig. 10(a)).

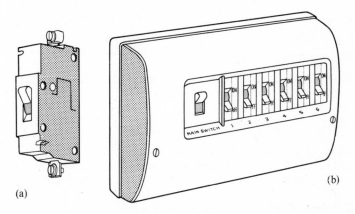

Fig. 10. *Excess-current circuit-breaker:* (a) *unit,* (b) *distribution board* (J. A. Crabtree & Co. Ltd.)

23

This device has several advantages over a fuse. These include: (*a*) all poles are disconnected in operation, (*b*) it can be tripped by remote control from pushbuttons, (*c*) after tripping, the circuit can be quickly reclosed, (*d*) the overload current and time delay may be adjustable.

The kind known as a *miniature circuit-breaker* (m.c.b.) can be incorporated in a distribution board, as in Fig. 10(b), in a similar manner to a fuse. Each circuit or subcircuit connected to a way of the board can thus be controlled and protected by a circuit-breaker. A main isolating switch may also be incorporated within the unit.

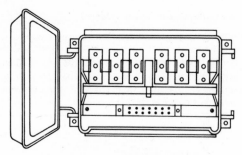

Fig. 11. Distribution fuseboard

Obviously, an m.c.b. board is more expensive than a corresponding distribution fuseboard, an example of which is given in Fig. 11.

Why must the neutral conductor be unfused?

In our standard 2-wire a.c. public electricity supply the neutral is earthed at the transformer substations. Consequently, on the development of a fault to earth, there is a low-resistance path through earth for the leakage current, so that the total current flowing will be sufficient to blow the fuse or trip the circuit-breaker. It is therefore essential for the neutral conductor to remain intact, as any break would nullify the system of excess-current protection.

What is discrimination?

Discrimination is the ability of a protective device, such as a fuse or circuit-breaker, to distinguish between excessive currents of different sizes. For instance, if, as the result of a fault on a subcircuit, a main fuse was caused to blow, this would indicate lack of discrimination. For proper protection, the subcircuit fuse, being of a lower rating than the main fuse, should have been the one to blow.

A mixture of different types of fuses on one installation often affects the discrimination. Most cartridge fuses offer satisfactory discrimination, provided that the same type is used throughout. Owing to the high cost of replacement of cartridge fuses, an installation is sometimes provided with cartridge fuses in the main fuseboard and semienclosed fuses in the subcircuit board. This means that discrimination cannot be obtained in the event of heavy excess-current.

When protection is provided by excess-current circuit-breakers, discrimination can be achieved by adjusting subcircuit breakers to operate not only at a lower current setting, but with a time lag shorter than that of the main breakers. For example, the subcircuit

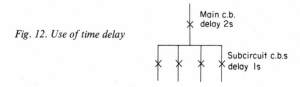

Fig. 12. Use of time delay

breakers could be set to operate 1 second and the main breakers 2 seconds after excess current started to flow, as shown in Fig. 12.

What precautions should be taken against mechanical damage?

We have seen that in most cases cable conductors and insulation require some form of protection against mechanical damage, in the form of a sheath, conduit, trunking or duct. In some situations, additional protection must be provided. Instances where this may be necessary are where cables are installed under floors or have to pass through walls, floors or structural metalwork. The additional protection may comprise tubing, bushing or bridging.

Nonmetal-sheathed cables exposed to direct sunlight (not through ordinary window glass), must be of a type having a p.v.c. or oil-resisting and flame-retarding (o.f.r.) or heat-resisting, oil-resisting and flame-retardant (h.o.f.r.) sheath, preferably black in colour.

Cables of lift installations, other than travelling cables, run in the shaft are required to be either armoured, metal-sheathed, enclosed in metal, o.f.r., or h.o.f.r.

Special care must be taken when handling p.v.c.- or paper-insulated cables during very cold weather, as these lose some of their flexibility under low-temperature conditions.

The minimum internal radii for bends in different kinds of cables for fixed wiring, and also the spacings of supports for cables in accessible positions, are tabulated in the I.E.E. Regulations.

What precautions should be taken against damage by heat?

Maximum ambient temperatures for cables and maximum normal operating temperatures for insulation and sheaths of cables are given in I.E.E. Regulations.

When conductors are included in a vertical channel, duct or trunking over 3 m in length, internal barriers must be provided to prevent the air at the top from reaching an excessively high temperature. For connections between ceiling roses and the lampholders in pendant fittings utilising tungsten-filament lamps, the use of heat-resisting flexible cords (e.g. those insulated with butyl rubber, ethylene propylene rubber, silicone rubber or glass fibre) is recommended. Similar precautions are recommended for batten lampholders in such situations.

It is stressed that exposure of plastics-insulated cables to high temperatures, even for short periods, may cause the insulation to soften.

Special precautions must be observed where cables are installed in situations liable to contain flammable and/or explosive materials.

What precautions should be taken against damp and corrosion?

Unless specially designed for the purpose, cables must not be installed where they will be exposed to water, oil or corrosive substances. Metal sheaths and armour of cables, metal conduit, trunking, ducts and their fixings installed in damp situations must be of corrosion-resisting material or finish. Such metal must not be placed in contact with other metals with which it is liable to set up electrolytic action.

Contact between bare aluminium sheaths or aluminium conduits and any parts made of brass or other metal having a high copper content must be especially avoided in damp situations, unless the parts are suitably plated. If such contact is unavoidable, the joint should be completely protected against the ingress of moisture.

Which circuits are required to be segregated?

For purposes of segregation, circuits are divided into three categories: (*a*) circuits (other than fire alarm circuits) operating at low or medium voltage and supplied directly from a mains supply system; (*b*) with the exception of fire alarm circuits, all extra-low voltage (not exceeding 30 V a.c. to earth) circuits, and telecommunication circuits (e.g. radio, telephone, sound-distribution, burglar alarm, bell and call circuits) which are not supplied directly from a mains supply system; (*c*) fire alarm circuits.

The I.E.E. Regulations lay down requirements regarding separation and partitioning of cables of the three circuit categories, the use of common enclosures and the mounting of controls or outlets on common boxes, etc., also the use of cores of multicore cables for different categories.

Other requirements refer to segregation from or bonding to exposed metalwork and other services.

What are the requirements regarding identification of conductors?

Every single-core cable and every core of a twin or multicore cable for use as fixed wiring must be identifiable at its terminations and preferably throughout its length. For ordinary p.v.c.-insulated

27

cables, core colours are used (e.g. red for live, black for neutral and green and yellow for earthing in the case of single-phase circuits). For armoured p.v.c.-insulated and paper-insulated cables, numbers for live cores and 0 for neutral may be used as an alternative to core colours. For mineral-insulated cables, sleeves or discs of appropriate colours are required at terminations.

Cores of flexible cables or cores should be identified as follows: live core(s)—brown; neutral cores—blue; earthing core—green and yellow. For p.v.c.-insulated parallel-twin nonsheathed flexible cords, the core having a longitudinal rib should be used for the live conductor.

What are the requirements for terminations?

The word *termination* may be applied to conduit, trunking, cable sheath or cable conductor. For example, a conduit can terminate in a metal box and a cable conductor can terminate in the terminal of an accessory. The principal requirement in connection with conduit terminations is that the ends are reamed and bushed (unless fitted into spout entries).

Cable conductor terminations are required to be mechanically and electrically sound. To fulfil the first requirement, the termination must be secure to the extent that it will not be dislodged in normal usage. To meet the second requirement there must be effective contact over as large a surface area as possible.

When terminating a stranded conductor in a soldered or compression type (crimped) socket, all of the wires must be securely contained and anchored.

Special care is needed in terminating aluminium conductors. Overtightening the clamping screw in a terminal, for example, may damage the conductor and lead to overheating.

Unsheathed cables at the termination of conduit, etc. must be contained in an incombustible enclosure, such as a box, accessory or lighting fitting.

Terminations of mineral-insulated and paper-insulated cables must be effectively sealed to prevent the ingress of moisture which would be absorbed by the insulation.

What are the requirements for joints?

Wherever possible, joints in conductors should be avoided. Where jointing is unavoidable, the requirements for terminations should, where appropriate, be observed.

Joints should be provided with insulation which is not less effective than that of the cores. Where practicable, they should be accessible for inspection.

3

WIRING SYSTEMS

How is a wiring system chosen?

Among the factors which may have to be taken into account in deciding the best system to use for a particular installation are the following:

(a) the conditions under which the work will have to be carried out (e.g. whether in occupied premises, on a construction site, etc.);

(b) any abnormal conditions to which the installation may be exposed (e.g. moisture, high or low temperatures, etc.);

(c) possibility of alterations or extensions;

(d) appearance (i.e. whether it can be on the surfaces of walls and ceilings or must be concealed);

(e) anticipated life of the building;

(f) time available for installation;

(g) cost.

Can bare conductors form a wiring system?

The I.E.E. Regulations recognise the installation of bare conductors in buildings for the following purposes:

(a) earth-continuity conductors, earthing leads and bonding leads (with certain exceptions);

(b) the external conductors of earthed concentric wiring;

(c) conductors of extra-low voltage wiring (provided that there is adequate insulation and further protection where necessary to ensure that the conductors do not cause risk of fire);

(d) protected rising-main and busbar installations (subject to certain conditions);

(*e*) collector wires for travelling cranes or trolleys or for similar purposes (subject to certain conditions).

How are p.v.c.-insulated p.v.c.-sheathed cables installed ?

Twin and earth or three-core and earth p.v.c.-insulated p.v.c.-sheathed cables (Fig. 13) can be installed without protection except where they pass through floors or walls or where they may be

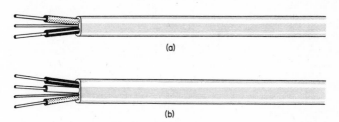

Fig. 13. *P.V.C.-insulated, p.v.c.-sheathed, 600/1000 V cable, flat with earth-continuity conductor:* (a) *two-core,* (b) *three-core* (British Insulated Callenders Cables Ltd.)

subject to mechanical damage (e.g. from heavy equipment). Advantages of such a system are that it is relatively cheap to install and can often be run unobtrusively.

Two alternative methods of wiring are commonly employed with this system. In one method, centrally placed joint boxes are used. The other method involves the use of a type of ceiling rose with an extra terminal.

Fig. 14 illustrates the joint box method. The subcircuit live and neutral conductors are each taken to terminals in the box and thence to lighting points and switches. Other terminals in the box are used for joints in the switch wires between switches and lighting points.

Fig. 15 illustrates the second method, in which the extra terminal in the ceiling rose is used to carry the live conductor through, the other terminals (from which the flexible cord is taken) being connected to switch wires and neutral. The terminal connected to the live conductor must be arranged (e.g. screened) so that it cannot be touched when the ceiling rose is dismantled to the extent necessary for replacement of the flexible cord.

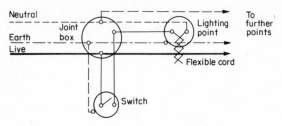

Fig. 14. Joint box method

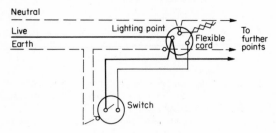

Fig. 15. Ceiling rose method

The cables of this particular system of wiring are usually fixed by buckle clips and saddles. It is good practice to secure single runs by clips and multiple runs by saddles. For fixing clips and saddles to most wall surfaces brass screws and plugs are suitable, but it is permissible to use brass pins for fixing direct to woodwork. Plastic cable clips incorporating specially hardened fixing pins are also used.

To secure cable by buckle clips the run is first marked out and the clips fixed at appropriate intervals. The maximum spacing in accessible positions for p.v.c.-sheathed cables varies from 250 mm horizontally and 400 mm vertically for the smaller sizes (not exceeding 9 mm overall diameter) to 400 mm horizontally and 55 mm vertically for the larger sizes (up to 40 mm overall diameter). The cable is laid across the fixing screw or pin then the ends of the clip are lifted, the tail is pushed through the eyepiece, drawn up tight and bent back on itself.

When securing by saddles, one screw or pin can be loosely fixed for each saddle in the run so that the cable or cables can be pulled taut and slipped into each saddle in turn and the fixing completed using the other screw or pin. In the case of multiple runs using metal saddles, it is possible to obtain strip or tape which can be cut, bent and drilled to suit the purpose.

How should cables be installed in conduits?

Conduits for mechanical protection of cables may be of: (*a*) steel, (*b*) nonferrous metal, (*c*) p.v.c.

The conduits for each circuit wired *in situ* must be completely erected before any cable is drawn in. There should be an adequate number of inspection type fittings (e.g. boxes, elbows, tees) installed so that they can remain accessible for such purposes as the withdrawal of existing cables or the installing of additional cables. The use of solid (noninspection) elbows or tees is restricted.

The maximum number of cables which can be run in one steel conduit is stipulated in the I.E.E. Regulations. The inner radius of

Fig. 16 Minimum inner radius of conduit bend

a conduit bend must not be less than 2·5 times the outside diameter of the conduit (see Fig. 16).

Ends of lengths of conduit must be reamed and where necessary bushed to prevent abrasion of cables. Unless intended to be gas-tight, conduit systems should be self-ventilating, and drainage

33

outlets must be provided at any points where condensed moisture might otherwise collect.

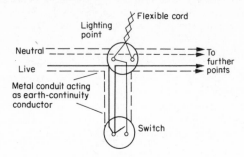

Fig. 17. Looping-in method

The method of wiring generally used for conduit systems is that known as looping-in, illustrated in Fig. 17, in which joints are made in the terminals of accessories.

What are the requirements regarding steel conduit systems?

Steel conduits used to contain cables are required to have a corrosion-resistant finish inside and outside. They must be effectively earthed. All joints must be mechanically and electronically continuous. This can be achieved by the use of mechanical clamps or by screwing. Subject to certain conditions, it can be utilised as the earth-continuity conductor. To avoid undesirable inductive effects, cables must be so arranged that the cables of all phases and the neutral conductor (if any) are contained in the same conduit.

Steel conduits may be bent using either a hand-bender or a conduit bending machine. Conduits may be cut by hacksaw and threaded using stocks and electric dies or, where a large number of joints are involved, a cutting and threading machine may be employed. Typical bends in steel conduits are shown in Fig. 18.

When steel conduit forming an earth-continuity conductor terminates at a metal box, it is essential that there is effective contact

34

between the conduit and the box. Two alternative methods of terminating at an unspouted box or a fuseboard, etc. are: (*a*) a female bush and locknut, and (*b*) a male bush and socket (see

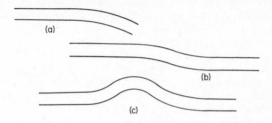

Fig. 18. Conduit bends: (a) *set,* (b) *double set,* (c) *saddle set*

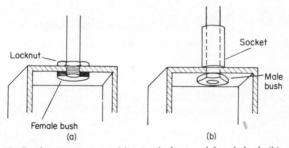

Fig. 19. Conduit terminations: (a) *using locknut and female bush,* (b) *using socket and male bush*

Fig. 19). The second method makes it easier to remove the box if it ever becomes necessary.

What are the requirements regarding nonferrous metal conduits?

Aluminium conduits are sometimes used to contain cables. In most situations this needs no additional protection such as painting. It will withstand, for instance, the sulphurous fumes associated with coal gas and producer gas. However, in damp situations the conduit should be painted with bituminous paint, for under these circumstances aluminium is liable to corrode when in contact with certain materials, particularly copper.

Since aluminium is softer than steel, it is more subject to mechanical damage and may require additional protection. On the other hand, it is lighter, easier to cut and thread, has better electrical conductivity and is nonmagnetic. Immediately after threading, the screwed conduit ends are smeared with petroleum jelly or other suitable grease as a protection against corrosion.

Zinc-base conduit may be used in a similar manner to aluminium conduit. Additional support may be necessary where such conduits are liable to be subjected to mechanical load or vibration. Care must be exercised in threading, as the material tends to tear.

In certain special situations copper conduits have been used in conjunction with bronze junction boxes and capillary type soldered fittings.

What are the requirements regarding nonmetallic conduits?

Nonmetallic conduits may be used only in situations where it is ensured that they are suitable for the extremes of ambient temperature to which they are likely to be subjected in service. For instance, rigid p.v.c. conduits may not be suitable for use where the normal working temperature of the conduits and fittings could be greater than 60°C or less than −5°C. Care must therefore be taken

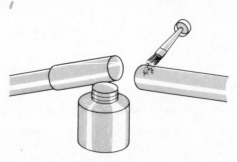

Fig. 20. Joining p.v.c. conduits using an adhesive (Gilflex Conduits Ltd.)

when a lighting fitting is suspended from a p.v.c. box that the temperature of the box does not exceed 60°C and that the weight of the fitting does not exceed 3 kg.

Rigid p.v.c. conduits should be supported in such a manner that longitudinal expansion and contraction of the conduits which may occur with variation of temperature under normal operating conditions is permitted. This type of p.v.c. conduit can be bent cold, although best results are obtained if it is warmed in a spirit lamp or gentle blowlamp flame to soften before bending. It can be joined by screwing or by the use of a suitable adhesive (Fig. 20).

A simple method of joining and terminating conduits, whether steel, aluminium or p.v.c., is that known as Easilock, in which conduits and fittings are push-fitted and locked by thread-forming screws (Fig. 21).

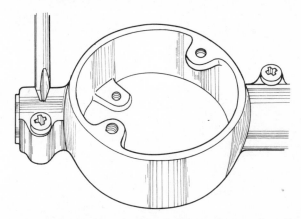

Fig. 21. Easilock method of joining and terminating conduits (Barton Conduits Ltd.)

Flexible type nonmetallic conduit, which is supplied in coils, enables the routing of conduits to be simplified, particularly when avoiding obstacles and negotiating awkward bends. Conduits comprising metal parts having an inner and outer covering of

37

insulating materials are regarded for purposes of the I.E.E. Regulations as nonmetallic conduits, provided that their metal parts cannot come into contact with other metal.

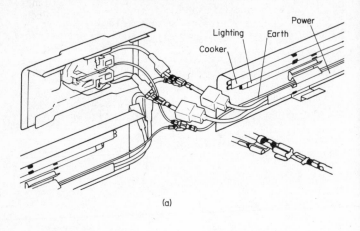

(a)

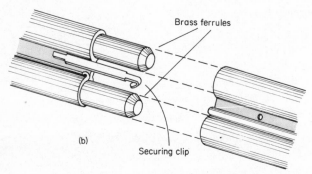

(b)

Fig. 22. Prefabricated wiring systems: (a) *Ampolex system using Faston type connections* (Aircraft-Marine Products (Great Britain) Ltd.), (b) *Octoflex system* (Hartley Electromotives Ltd.)

What are the requirements regarding prefabricated conduits?

This type of system is not wired *in situ*, but usually consists of a harness or conduit assembly prewired in a factory and delivered to the site for fixing. It is especially applicable to system building, where there are a number of similar units, such as flats, offices, hospital wards, school classrooms, etc.

In one such system, cables are contained in two or three semi-rigid p.v.c. tubes joined by p.v.c. webbing which provides a means of fixing. Either a consumer's control unit or a miniature-circuit-breaker panel is included in the wiring kit with necessary accessories for fixing and connecting in position.

It is important that adequate allowance is made for any possible variations in building dimensions, so that conduits or cables are not subjected to tension or other stress during installation. Precautions must be taken against damage due to building operations, especially against deformation of conduits and damage to exposed cable ends.

In industrialised building construction it is an advantage to have an electrical system which enables the wiring to be carried out independently of the main structure of the building via skirtings, architraves and cornices. One method (Ampolex, see Fig. 22(a)) utilises plastics-covered metal skirting and architrave sections with purpose-designed accessories and quick connections. Yet another method (Octoflex, Fig. 22(b)) uses cables contained in 2- or 3-bore webbed tubes.

The 'harness' system is undoubtedly the quickest method, but in many cases, owing to the nature of the building or other features, it cannot be employed.

What are the requirements regarding duct and trunking systems?

A cable *duct* is defined as a 'closed passage-way formed underground or in a structure and intended to receive one or more cables which may be drawn in.' A cable *trunking* is defined as 'a fabricated casing for cables, normally of rectangular cross-section, of which one side is removable or hinged to allow cables to be laid therein.'

Types of duct used to contain cables vary from those formed of trapeziform section p.v.c. to those comprising large purpose-built passages. Trunking may be of steel or noncombustible insulating material of various sizes and sections.

Many of the I.E.E. Regulations referring to conduit systems also apply to those utilising trunking. However, the maximum space factor for trunking is 45 per cent and that for ducts 35 per cent.

What is the underfloor duct system?

It is a system applicable to concealed electrical distribution in offices, workshops, supermarkets, laboratories and similar areas. It is primarily intended for installation within floor screeds and provides quick and easy access to electrical services without structural alterations or damage to decorations.

A typical system of this kind (Fig. 23(a)) involves the use of ducting extruded from grey rigid p.v.c. in trapeziform section of standard sizes (e.g. 60 mm and 90 mm). The ducting is normally laid out on a grid or comb pattern of single, double or triple runs. Modular components for intersections and corners are included in the range of standard fittings, and concealed outlet boxes are provided beneath the floor surface.

What is the skirting duct system?

This system, as the name implies, consists of a skirting specially designed to contain electrical wiring (Fig. 23(b)). It may be used as a complete system or be complementary to an underfloor duct system.

What is the Ductube system?

As shown in Fig. 24, it is a system of wiring in ducts formed in concrete by means of inflatable rubber tubes. The tubing is placed along the wiring routes and inflated before the concrete is poured. When the concrete is set, the tubing is deflated and withdrawn so as to leave a circular bore to receive the cables.

The radial thickness of concrete or screed surrounding the cross-section of a completed duct must not be less than 15 mm at every

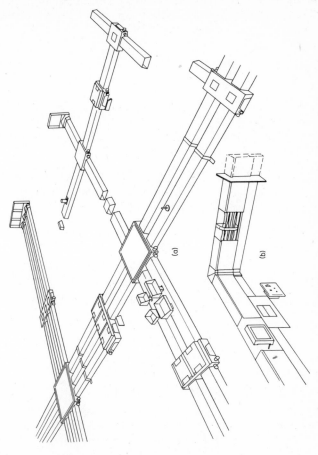

(a)

(b)

Fig. 23. Duct systems: (a) underfloor duct, (b) skirting duct (Key Terrain Ltd.)

point. The system is not applicable to nonsheathed p.v.c.-insulated cables or rubber-insulated braided and compounded cables.

Formers are used for ceiling fittings, switches and socket-outlets.

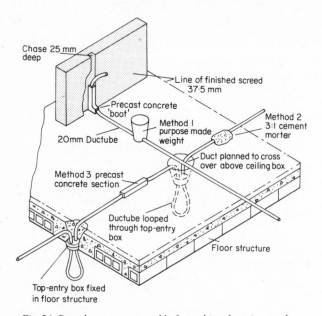

Fig. 24. Ductube system: assembly for making ducts in screed.
(Ductube Ltd.)

What is the overhead busbar trunking system?

This system, shown in Fig. 25, is particularly suited to situations where there are lines of machines or other current-using equipment. The usual arrangement comprises rectangular section trunking containing copper or aluminium busbars. At intervals, say, every metre, a tapping-off connection point is provided. This may include fuses to protect the subcircuits. Connections to the items of equipment could be by p.v.c.-insulated cables in conduit, armoured p.v.c.-sheathed cables or by mineral-insulated cables.

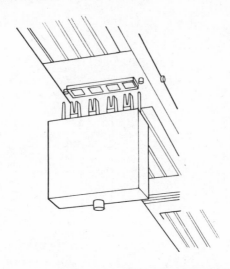

Fig. 25. Overhead busbar system (G.E.C.-English Electric Ltd.)

This system enables much of the electrical installation work to be carried out before the precise position of the equipment is determined. It also facilitates subsequent additions and alterations.

What is the rising-main busbar system?

It is a system often used for distribution in multistorey blocks of offices or flats. Usually, copper or aluminium busbars are supported by insulating supports within a trunking. Fire barriers are positioned at the top of each fixed section of trunking passing through floors, and the busbars are sleeved with p.v.c. at these points (Fig. 26).

Isolators or switch-fuses may be mounted on the front of the rising trunking at appropriate positions (e.g. for each floor). Alternatively, the switchgear may be mounted alongside the rising mains with a short spur of reduced size trunking from the side to house the connecting wiring.

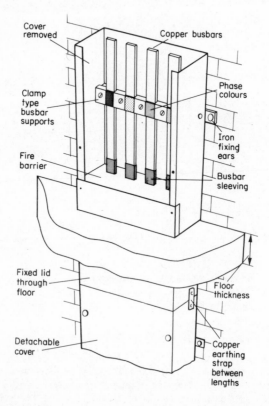

Fig. 26. Rising-main system (Davis Sheet Metal Engineering Co. Ltd.)

What are the requirements regarding paper-insulated lead-sheathed and armoured cables?

These are mainly used for heavy current work, such as mains and submains. The construction of typical cables of this type suitable for 600/1000 V, 3-phase 4-wire supplies is shown in Fig. 27, which illustrates steel tape armoured cables having: (*a*) stranded conductor, and (*b*) solid aluminium conductor. In the larger sizes, the cable is provided with either a belt of paper tape applied overall or a layer of metallised paper tape or metal tape on each core, the laid-up cores being bound together with copper woven fabric tape. To give maximum flexibility, steel wires are used for armouring instead of steel tapes.

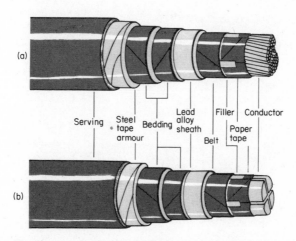

Fig. 27. Paper-insulated, lead-sheathed, armoured cable: (a) *stranded conductor,* (b) *solid aluminium conductor* (British Insulated Callenders Cables Ltd.)

The cables may be installed indoors or outdoors and either laid directly in the ground or in ducts. Long runs are normally buried directly and protected by tiles where they are liable to be damaged (e.g. by picks or shovels). In some situations the cables are drawn into earthenware ducts.

Since the impregnated paper absorbs moisture when exposed to the atmosphere, all cable terminations must be effectively sealed.

What are the requirements regarding p.v.c.-insulated armoured p.v.c.-sheathed cables?

These are to some extent superseding paper-insulated cables for mains and submains. They may consist, as shown in Fig. 28, of solid or stranded conductors surrounded by p.v.c. insulation, bedding, galvanised steel wire armour and p.v.c. oversheath.

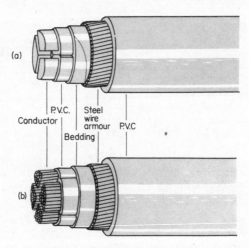

Fig. 28. P.V.C.-insulated, p.v.c.-sheathed, armoured cable: (a) *solid aluminium conductor,* (b) *stranded conductor* (British Insulated Callenders Cables Ltd.)

This type of cable is relatively simple to terminate and joint using compression type glands. It has the advantages of immunity to damage by moisture and inherent toughness with flexibility over a

46

wide temperature range. It is available with conductors of either solid aluminium or stranded copper.

The minimum permissible internal radius of a bend in this cable is eight times the overall diameter. The maximum spacing of supports in accessible positions for cables up to 15 mm overall diameter is 350 mm horizontally and 450 mm vertically.

What are the requirements regarding mineral-insulated metal-sheathed cables?

Mineral-insulated metal-sheathed cables commonly used include those having copper conductors and copper sheath, copper conductors and aluminium sheath, and aluminium conductors with aluminium sheath. All can be obtained with a p.v.c. oversheath.

In common with paper insulation, the mineral insulation absorbs moisture when exposed to the atmosphere, which means that all terminations have to be effectively sealed. This is accomplished generally using a pot containing plastic compound. A typical seal for use in temperatures up to 80°C is shown in Fig. 29. The seal for temperatures up to 105°C utilises a brass pot shell in place of the plastics one illustrated. When glass fibre or metal discs and p.t.f.e. sleeves are used, the permissible operating temperature is increased to 135°C with ordinary compound and 185°C with special compound.

Advantages of this type of cable include: (a) nonflammable, non-ageing insulation, (b) when correctly installed the system is impervious to water, oil and many liquids, (c) appreciable mechanical strength, (d) relatively high current rating, (e) small overall diameter.

Fixing of the cable to surfaces is by means of copper clips or saddles; alternatively, it can be run on brackets, hangers, racks or trays. If fitted with a p.v.c. oversheath, it can be buried directly in the ground.

The minimum permissible radius of a bend is six times the overall diameter. The maximum spacing of supports in accessible positions is 600 mm horizontally and 800 mm vertically for cables up to 9 mm overall diameter, and 900 mm horizontally and 1200 mm vertically for cables over 9 mm and up to 15 mm overall diameter.

Certain plasters and cements cause corrosion of copper, so that it is advisable to know the constitution of such materials before

47

burying cables directly in them. Unprotected cables should not be buried beneath terazzo finishes. However, a p.v.c. oversheath provides almost complete protection against corrosion.

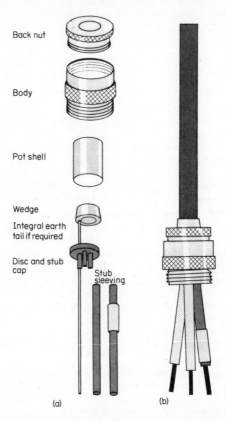

Back nut

Body

Pot shell

Wedge

Integral earth tail if required

Disc and stub cap

Stub sleeving

(a)

(b)

Fig. 29. Terminations of mineral-insulated, copper-sheathed cable: (a) *accessories,* (b) *completed termination* (British Insulated Callenders Cables Ltd.)

Particular care is required in connection with protecting aluminium against corrosion. For example, aluminium should never be in direct contact with brass or any other metal having a high copper content unless suitable precautions are taken to prevent corrosive action that may occur, particularly under damp conditions.

What is earthed concentric wiring?

It is a system employing cables with concentric conductors. Thus, it applies to the installation of metal-sheathed cables using the sheath as the conductor connected to an earthed neutral. Although it does have much to recommend it, it tends to complicate terminating

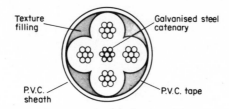

Fig. 30. Catenary cable (Enfield Standard Power Cables Ltd.)

where it entails connections to the sheath. Moreover, there are potential dangers arising from the possibility of the development of an open-circuit or high-resistance fault in the sheath. The system can only be used when:

(*a*) the supply undertaking concerned has been specially authorised in respect of the installation by the Minister of Technology or the Secretary of State for Scotland (or other relevant authority) to permit additional connections of the neutral conductor to earth, or

(*b*) where it is supplied by a transformer or convertor in such a manner that there is no metallic connection with a public supply, or

(*c*) where the supply is obtained from a private generating plant.

The general conductor, which must be earthed, may not be common to more than one circuit, although twin or multicore cables may serve a number of points in one final subcircuit. The resistance of the external conductor should not be greater than that of the internal conductor or (if appropriate) the internal conductors in parallel. Any joints in the external conductor should be in addition to any means used for sealing and clamping. No fuse, non-linked switch or circuit-breaker must be connected in the external conductor.

What is catenary wiring?

This is suitable for wiring within high-ceilinged buildings or across spaces between buildings. In a basic catenary system, insulated cables are suspended on insulated hangers from a steel wire stretched between opposite walls or other supports. A manufactured version (Fig. 30) utilises an arrangement of insulated cables made up round a high tensile galvanised steel wire, packings of jute or hemp being used to produce a circular section. The assembly is installed across the required space by means of the steel wire, which is secured at each end by eye-bolts and strainers.

A standard form of connection box may be used for either tee, angle or through joints. The box may also be used to accommodate fuses and neutral link, and possibly as a point of suspension.

Made-up catenary wiring can generally be suspended across spans up to about 50 metres, but it is usual to provide auxiliary suspension at more frequent intervals. Sometimes, the complete span assembly, including fittings, boxes, etc., is assembled on the ground before erecting.

4

LIGHTING

Why are most electric lamps connected in parallel?

Items of electric equipment can be connected to the electricity supply in either *series* or *parallel*. The connection of ordinary lamps in series as in Fig. 31(a) would have the following disadvantages:

(*a*) unless each lamp incorporated some form of automatic shorting device, failure of one lamp would cause all of the lamps in the subcircuit to be disconnected;

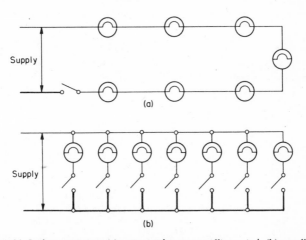

Fig. 31. Lighting points in: (a) *series, with one controlling switch,* (b) *parallel, with each point individually controlled*

(*b*) it would not be practicable to control the lamps separately;

(*c*) to permit variation in the number of lamps connected in different subcircuits, lamps suitable for many different voltages would have to be available, as the mains voltage would be shared between all of the lamps.

Thus, lamps for general service lighting are connected in parallel as in Fig. 31(b), enabling each to operate and to be controlled independently and to be designed for the voltage of the public supply.

Christmas decoration lamps are usually connected in series because they are supplied in sets of a definite number, each designed to be operated on a very low voltage obtained by dividing the supply voltage by the number of lamps in the set. For example, if there were 20 lamps in series, each lamp would be designed for

$$\frac{240}{20} = 12 \text{ volts}$$

What is a single-pole switch?

A single-pole switch (Fig. 32(a)) is one which is designed to break, or disconnect, one pole only of the supply. A single-pole lighting switch must always be connected in the live pole of the supply; never in the neutral. Although a single-pole switch in the neutral

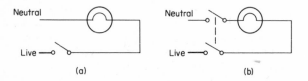

Fig. 32. Switches: (a) *single-pole,* (b) *double-pole*

conductor would allow the lamp to be switched on and off, it would leave one terminal of the lighting point permanently live, and could result in a shock to someone renewing a flexible cord after switching off at the lighting switch.

A double-pole switch (Fig. 32(b)) disconnects both poles of the supply and is often used as a main switch or to control heavier current apparatus.

What is one-way control?

It is the commonest kind of lighting control in which a switch is used to switch on or switch off one or more lighting points from one position (Fig. 33(a)). Only a limited form of control is provided by the use of two one-way switches. The bedroom light in Fig. 33(b),

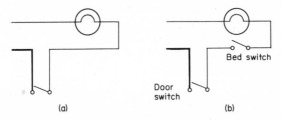

Fig. 33. (a) *Ordinary one-way control from one position,* (b) *limited control from two positions*

for instance, could be controlled by one switch at the bedroom door and another switch by the bed, provided that the user remembers each morning to restore the bed switch to the on position and the door switch to the off position. The point is that both switches must be on before the light will operate.

What is two-way control?

This is a method in which two two-way switches are used to control one or more lighting points from each of two positions inde-pendently. This involves the provision of cables known as *strappers*

53

(Fig. 34(a)). An alternative method of connecting two two-way switches is shown in Fig. 34(b).

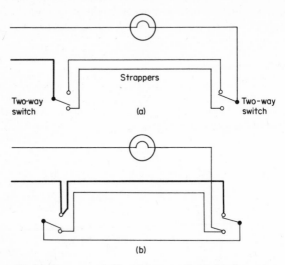

Fig. 34. Two-way switching: (a) *conventional,* (b) *alternative*

What is two-way and intermediate control?

This is a method of switching which is used when it is necessary to control one or more lighting points independently from three or more positions. The intermediate switches, having four terminals, are connected between two two-way switches as shown in Fig. 35. Any number of intermediate switches can be included between a pair of two-way switches.

How is a ceiling rose used?

A ceiling rose is generally fitted at the termination of a lighting point. Its purpose is to provide a safe and convenient method of suspending a flexible cord pendant and of joining it to conductors

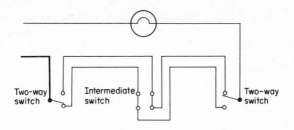

Fig. 35. Two-way and intermediate switching

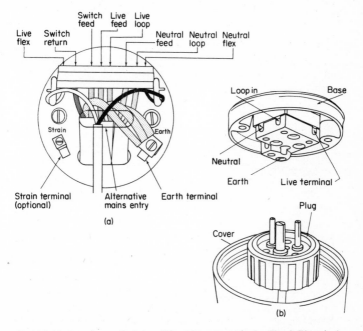

Fig. 36. (a) *A modern ceiling rose,* (b) *ceiling rose with plug* (Rock Electrical Accessories Ltd.)

of the fixed wiring. It consists basically of terminals mounted in or on an insulating base provided with an insulating cover and a grip for the flexible cord.

Ceiling roses may have three or four terminals, including the earth connection. If one of the terminals is intended to be used for looping to other points, the terminal is shrouded with insulating material. A modern design of ceiling rose is illustrated in Fig. 36(a).

One design of ceiling rose (Fig. 36(b)) incorporates a plug which enables individual lighting fittings to be safely and easily disconnected without interrupting other points in the subcircuit. Unless specially designed for the purpose, a ceiling rose must not be used for the attachment of more than one outgoing flexible cord.

When is a ceiling switch used?

A ceiling switch, or cord-pull switch, is used in a bathroom for purposes of safety, or in two-way switching in other rooms (e.g. bedrooms) for convenience (Fig. 37). It is also suitable for use

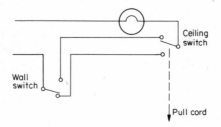

Fig. 37. Ceiling switch in bedroom

where the existence of a large window or a glass partition prevents the fixing of a wall switch in a suitable position. There are also circumstances where a ceiling switch can be used instead of a wall switch to simplify installation and to avoid the necessity to chase brickwork.

How are lampholders used?

These are, naturally, designed to hold lamps and to keep them in contact with the subcircuit conductors. Mains voltage filament lamps up to and including 150 W rating are housed in bayonet contact (b.c.) lampholders. Lamps of higher rating are designed to be accommodated in screwed lampholders.

The I.E.E. Regulations require that every b.c. lampholder in a damp situation or in any situation where it can be readily touched by a person in contact with earthed metal must be either earthed, shrouded in insulating material or fitted with a protective shield. In a bathroom, a protective shield is required to prevent the touching of parts of a lampholder by a person replacing a lamp.

Centre-contact bayonet or Edison-type screw lampholders have their outer or screwed contacts connected to the earthed neutral conductor of the supply. In the absence of an earthed conductor or in a damp situation, such lampholders must have a protective insulating shield.

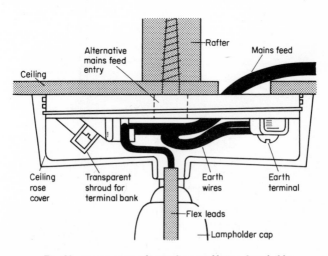

Fig. 38. Arrangement of 'swivel' type of batten lampholder

It is recommended that heat-resisting flexible cords be used between the ceiling rose and lampholder in pendant lighting fittings and in certain other types of fittings.

Batten lampholders are used where it is required to fit them direct to surfaces of ceilings and walls. In the 'swivel' type of batten lampholder (Fig. 38), permanent cables are housed in a separate ceiling rose or joint box to minimise heat transfer. Heat-resisting p.v.c., glass fibre or silicone rubber tails are fitted in conditions of high ambient temperatures.

How are joint boxes used?

A joint box, as the name implies, is designed to contain joints in conductors. The design varies according to the system for which the joint box is to be used. A joint box for p.v.c.-sheathed cables, for instance, is generally a moulded box of insulating material

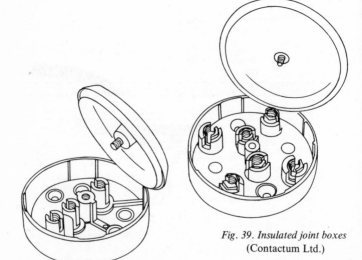

Fig. 39. Insulated joint boxes
(Contactum Ltd.)

58

inside which is the required number of shrouded terminals. The box has knockout entries and is provided with a lid (Fig. 39).

For use in conjunction with metal-sheathed and/or armoured cables, there are specially designed metal joint boxes. If non-metallic joint boxes are used for these cables, there must be a method of maintaining earth continuity. A bonding strip used for this purpose must have a resistance no higher than that of an equivalent length of earth-continuity conductor.

How does a filament lamp operate?

In a modern general purpose filament lamp (Fig. 40(a)), a fine tungsten wire sealed within a glass bulb is heated to incandescence (white heat) by the passage of an electric current. A gasfilled lamp contains a mixture of argon and nitrogen to slow down evaporation from the filament which is closely coiled to concentrate the heat. The higher the temperature at which the filament operates, the higher is the efficiency of light output.

In the type of filament lamp known as a tungsten iodine lamp (Fig. 40(b)), which is used in floodlighting, evaporation from the filament is controlled by the presence of iodine vapour.

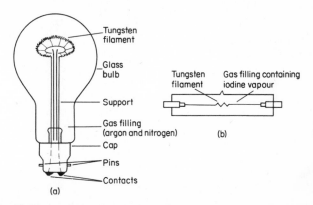

Fig. 40. Filament lamps: (a) *ordinary gasfilled tungsten filament,* (b) *tungsten iodine*

How does a discharge lamp differ from a filament lamp?

Discharge lamps of the mercury or sodium vapour type do not have filaments. The electric current is conducted through inert gas or vapour sealed within the glass envelope. This is quite different from the function of the gas within a filament lamp which takes no part in the production of light.

Unless something were included to prevent it, vaporisation would increase as more and more heat was produced until breakdown occurred. The device used to stabilise the discharge, sometimes called a *ballast*, may be a choke, transformer or resistor.

How do mercury-vapour lamps operate?

A high-pressure mercury-vapour lamp (Fig. 41(a)) has an inner quartz tube (the arc tube) containing a blob of mercury together

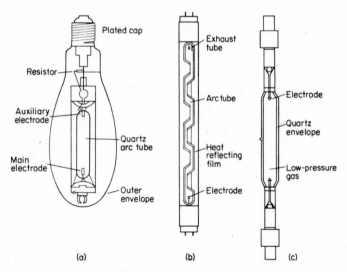

Fig. 41. Discharge lamp: (a) *high-pressure mercury-vapour lamp,* (b) *linear sodium-vapour lamp,* (c) *linear xenon lamp*

with an inert gas or gases at pressures of from 1 to 10 atmospheres according to type. Metal halides are often included. The arc tube is surrounded by an outer protective glass envelope.

The discharge is initiated across the gap between an auxiliary electrode and a main electrode in the high-pressure type of lamp. As the temperature rises, mercury is vaporised and the discharge takes place between the two main electrodes, since the resistance of this path is lower than that through the resistor in series with the auxiliary electrode.

Why after switching off do these lamps fail to restart immediately?

Because, due to the vaporisation of mercury, the internal pressure is too high for starting. Therefore, a lamp can only be restarted when sufficient mercury has recondensed, which in some types takes a matter of several minutes. Under these circumstances, no damage will be done to the lamp by leaving it switched on while not operating.

What is a compact source lamp?

This is a type of mercury-vapour lamp of small size and high brightness. The quartz arc tube is either spherical or isothermal in shape and the internal pressure is high, usually more than 20 atmospheres.

There are several types of compact source lamps. One type has an outer glass envelope; another type has no outer envelope; yet another type is contained within a metal box with a quartz window.

Compact source lamps are used principally in projection, photo-micrography and fluorescent microscopy.

What is a black glass lamp?

It is a mercury vapour lamp having an envelope of black 'Wood's' glass which absorbs the visible radiation and transmits ultraviolet radiation of 365 mm wavelength suitable for exciting fluorescent materials.

This kind of lamp is used in operations where marks invisible in normal lighting are required to be seen and in the detection of spurious matter (e.g. in food processing inspection) and in the detection of forgeries in documents, paintings, stamps, etc. It is also employed in stage shows and advertising displays for colourful effects using fluorescent powders, paints and dyes.

How do sodium-vapour lamps operate?

Instead of mercury, these lamps (Fig. 41(b)) contain, in addition to inert gas, small quantities of sodium, which results in a discharge of the characteristic yellow light which kills most colours, but is very efficient from a visual point of view and particularly effective under foggy conditions when used for street lighting.

There are several types, some having U-shaped arc tubes with removable outer jackets, some having U-shaped arc tubes with integral jackets, and some having indented kidney section arc tubes with linear arrangement and connecting pins at each end. In general, sodium lamps must be operated in a principally horizontal position.

Sodium has a very low vapour pressure, and for efficient light production requires an arc tube temperature of about 270°C. Special sodium-resistant glass is necessary for the arc tubes.

Why do discharge lamps result in a low power factor?

When a choke or transformer is included in an a.c. circuit, some of the power, known as 'wattless' power, is utilised in magnetisation of the iron cores, and this would not be recorded on a wattmeter connected in the circuit. Thus, the watts are less than the product of volts and amperes. This condition of *low power factor* results in current which lags behind the voltage and has the disadvantage of requiring equipment of a higher rating than is actually necessary for the work in question.

How is the power factor improved?

The low power factor is improved by connecting a capacitor in parallel with the supply to the discharge lamp concerned. This has the effect of introducing current which leads the voltage and so neutralises some of the lagging current due to the low power factor.

What is stroboscopic effect?

This is the effect that causes light from a discharge lamp to flicker due to cyclic variation. It is a disadvantage in a situation where there is moving machinery. For instance, it is possible that, due to the stroboscopic effect, a rotating part may appear to be stationary.

The effect can be minimised by circuit design.

What is the principle of a hot-cathode fluorescent lamp?

This is a type of discharge lamp which does not rely for light production directly on the glow resulting from the discharge of electricity through the gas or vapour within the tube. In this case, the discharge is designed to produce ultraviolet radiation suitable for activating the fluorescent material with which the inside of the glass tube is coated.

What is switch-start fluorescent lighting?

In switch-start lamps, the cathodes, which are the metal electrodes at the ends of the tubes, are fitted with heaters to stimulate the emission of electrons necessary for the discharge. A starter switch is included in the circuit to provide in conjunction with the choke the momentary interruption necessary to create a voltage surge, or pulse, for starting the discharge.

The starter switch therefore delays application of the starting voltage across the lamp until the electrodes have been heated to their operating temperature. The voltage pulse for starting generally occurs after 1 or 2 seconds.

How do starter switches operate?

There are two kinds of starter switches: (a) the thermal type, having two contacts and a heater (Fig. 42(a)), and (b) the glow type, having two contacts in a gas (Fig. 42(b)). In both types one of the contacts is on a bimetallic strip which bends with appreciable change in temperature.

On switching on the circuit containing a thermal starter, current flows across the closed starter switch contacts and the lamp electrodes are heated. After a short period, the heat within the starter is sufficient to cause the bimetallic strip to bend and to break the circuit. This gives the high-voltage pulse to start the discharge. While the lamp is operating, the heater keeps the start-switch contacts apart.

In a glow starter switch, the contacts are apart on switching on, and a glow discharge occurs between them. The heat thus produced causes the bimetallic strip to bend and the two contacts to touch. Current then flows, and the lamp electrodes are heated to operating temperature. When the bimetallic strip cools, it bends, breaks the circuit, and causes the pulse necessary to start the discharge.

What is switchless-start fluorescent lighting ?

There are ways of avoiding the preheating of the cathode and the consequent need for a starter switch. In one method (Fig. 42(c)), the fluorescent lamp either has an external earthed conductive coating or is arranged to be housed close to earthed metalwork. An autotransformer is used as the ballast instead of a choke. The discharge is initiated between the cathodes and the external earth connection.

Another method utilises lamps each having an internal conductive strip connected to one cathode and extending almost to the other cathode, the initial discharge taking place across the gap. The ballast in this case is a filament lamp.

What is the two-lamp series circuit

This is a switchless start circuit (Fig. 42(d)) in which two short fluorescent lamps are connected in series. The transformer used has three secondary windings, but the two outer ones can be made continuous with the primary winding.

What is the lead-lag circuit?

It is a method (Fig. 42(e)) of connecting two fluorescent lamps to reduce the flicker or stroboscopic effect. One lamp is controlled by a choke only, and the other lamp by a choke and capacitor in series. This causes the light flicker of one lamp to be out of step with that

64

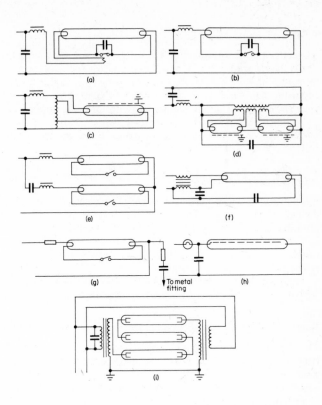

Fig. 42. Fluorescent lighting circuits: (a) *thermal switch start,* (b) *glow switch start,* (c) *switchless start,* (d) *two-lamp series,* (e) *lead-lag,* (f) *semi-resonant,* (g) *resistor ballast,* (h) *filament lamp ballast,* (i) *three-lamp cold-cathode*

of the other lamp, making the resultant light level much steadier than it would otherwise be.

What is a semi-resonant circuit?

It is a circuit incorporating (Fig. 42(f)) a transformer and a capacitor across the electrodes. It is designed to produce a degree of electrical resonance after switching on, so that a sufficiently high voltage is applied to start the discharge, the electrodes having been preheated. Once the discharge has started, the capacitor provides power factor correction. The fluorescent lamp used has a conducting device (metal strip or silicone coating) to reduce the necessary starting voltage.

What is the resistor ballast circuit?

It is a circuit in which the choke is replaced by a resistor to control the current (Fig. 42(g)). This means that there is no voltage pulse resulting from a break in the circuit on starting. Instead, the fluorescent lamp is silicone-coated and provided with an artificial earth to the metalwork of the fitting via a resistor and capacitor in series.

Although the efficiency is appreciably less than that of a choke control circuit, the power factor is high (slightly below unity).

What is the filament lamp ballast circuit?

This is a circuit (Fig. 42(h)) in which a particular type of filament lamp is used as the ballast and the fluorescent lamp has a high-resistance internal metal strip extending from one electrode to within a short distance of the other (as mentioned under the heading of switchless-start fluorescent lighting).

On switching on, an arc strikes across the gap between the electrode and the end of the strip. This creates the heat necessary to promote conditions for starting the main discharge. The filament lamp contributes slightly to the light output. To prevent it from being connected directly to the mains voltage in error, it is fitted with a three-pin bayonet cap.

Can fluorescent lamps be dimmed?

By use of a suitably designed circuit, a fluorescent lamp can be dimmed. One such circuit utilises in addition to a choke a special transformer in which two secondary windings supply full heating current to the lamp electrodes, irrespective of the position of the dimming device. The actual dimming may be achieved by a variable resistor.

Can fluorescent lamps be operated on direct current?

There are inherent disadvantages in operating a fluorescent lamp from a d.c. supply. In addition to the use of a choke for providing the starting voltage pulse, a resistor must be connected in series to limit the lamp current to the correct value. A thermal starter switch or a specially designed glow starter switch is also required.

In addition, a polarity-reversing switch must also be included in the circuit to reverse the polarity each time the circuit is switched on. This is necessary to prevent the continual migration of mercury to one electrode (the negative) that would otherwise occur.

What is cold-cathode fluorescent lighting?

This form of lighting generally utilises long fluorescent tubes of 22 mm diameter with very low internal pressure. The cathodes are thin Swedish iron cylinders and are not preheated. The tubes are of the type employed in neon signs and displays, and are operated via high-voltage transformers which entail special safety measures at points where there may be access to high voltage.

A typical cold-cathode lighting fitting comprises three straight tubes about 2·5 m long operated by two transformers in tandem (Fig. 42(i)). There are also fittings containing curved tubes. Cold-cathode fluorescent tubing has a very long life and is thus often installed in situations where it is not readily accessible.

What are xenon lamps?

They are lamps containing the rare gas xenon, the electrical discharge through which produces light very akin to sunlight. They can operate over a wide range of pressures, and with very high loadings. No starter switch is required.

There are two main types of xenon lamp: compact source (used for projection), and linear (Fig. 41(c)) (used in the smaller sizes for fadeometers, and in the larger sizes for floodlighting). Both the lamps and the control gear are very expensive.

What are electroluminescent lamps?

These are light sources in which light is produced by materials called phosphors when exposed to electric fields. Phosphors commonly employed are sulphides or sulphoselenides of zinc or cadmium.

The lamps are usually in the form of luminous panels or strips which vary in construction according to type. However, the general arrangement incorporates activation of the phosphor by an electrically conducting plate connected to one pole of the supply. The plate has a thin layer of phosphor embedded in resin or ceramic. Above this layer there is a glass plate on the underside of which a transparent conducting film (e.g. tin oxide) is deposited. The phosphor acts as the dielectric of a capacitor.

Although the luminous efficiency of such a device is low, it is suitable for luminous surrounds to signs, clocks, bell pushes, nightlights, etc.

5

SOCKET-OUTLETS, HEATING AND COOKING

What is the purpose of socket-outlets?

A socket-outlet is defined as 'a device with protected current-carrying contacts intended to be mounted in a fixed position and permanently connected to the fixed wiring of an electrical installation to enable the connection to it of a flexible cord or flexible cable by means of a plug.'

Socket-outlets are widely used for connection of portable electrical appliances, such as reflector fires, television receivers and hand drills, to the supply. The term 'appliance' as used in the I.E.E. Regulations is any device which utilises electricity for a particular purpose, excluding a lighting fitting or an independent motor. An appliance intended to be fixed to a supporting surface or to be used in any one place is known as a 'stationary appliance'.

Socket-outlets are sometimes referred to as plug points and often, wrongly, as power points. The actual sockets are designed to receive the pins of the plugs to which items of equipment are connected by means of flexible cords.

Which different kinds of plugs and socket-outlets are acceptable?

Types of plugs and socket-outlets acceptable for use on low-voltage circuits (those on which the voltage does not exceed 250 V) include the following:

Domestic and commercial type 2-pole and earth plugs (fused and unfused) and socket-outlets of 2, 5, 15 and 30 A rating;

Protected type 2-pole with earthing contact plugs (fused and unfused) and socket-outlets of 5, 15 and 30 A rating;

Theatre type plugs and socket-outlets of 15 A rating;

Industrial type plugs and socket-outlets of 16, 32, 63 and 125 A rating;

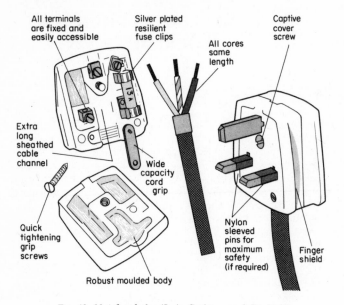

Fig. 43. 13 A fused plug (J. A. Crabtree and Co. Ltd.)

Fused plugs (Fig. 43) (with fuses rated at 3 and 13 A), 2-pole and earth, and shuttered socket-outlets rated at 13 A.

What are the requirements of socket-outlets and plugs?

The I.E.E. Regulations require that it shall not be possible for any pin of a plug to be engaged with any live contact of its associated socket-outlet connected to mains voltage while any other pin of the plug is completely exposed. Also, it must not be possible for any pin of a plug to be engaged with any live contact of a socket-outlet within the same installation other than the type of socket-outlet for which it is designed. Every plug containing a fuse must be non-reversible and so designed and arranged that no fuse can be connected in an earthed conductor.

For the standard 2-wire a.c. supply, socket-outlets must be connected so that the live socket (having a terminal marked L) is on the right-hand side when facing the socket-outlet from the front, the neutral being connected to the terminal marked N on the left-hand side and the earth to the terminal marked E above centre. The colour identification of the cables of the fixed wiring is red for live, black for neutral and green and yellow for earth.

Socket-outlets mounted on a wall should be at a minimum height of 150 mm from the floor level, or 150 mm above a working surface (e.g. in a kitchen).

The plugs should be arranged so that their pins engage with the corresponding sockets in the socket-outlet. The colours of the cores of the flexible cord connected to a plug should be brown for live, blue for neutral and green and yellow for earth.

What is a radial final subcircuit?

It is one in which the outlets are connected in parallel with the supply terminals by single conductors (Fig. 44). The I.E.E. Regulations require that each final subcircuit be connected to a separate way in the distribution board and that the wiring of each final subcircuit be electrically separate from that of every other final subcircuit.

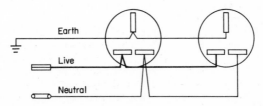

Fig. 44. Radial final subcircuit

71

The number of socket-outlets served by a radial final subcircuit of rating not exceeding 15 A is limited by their aggregate demand. For socket-outlets rated at 5, 13 or 15 A, the current demand to be assumed is 5, 13 or 15 A respectively. For a 2 A socket-outlet, a demand of at least 0·5 A must be assumed.

Socket-outlets connected to radial final subcircuits may be of the round pin variety with unfused plugs. A distinction is made in the I.E.E. Regulations between the requirements for socket-outlets on domestic and nondomestic circuits.

What is a ring final subcircuit?

This is a subcircuit in which each conductor is run in the form of a ring, commencing from a way in a distribution board (or its equivalent), looping into the terminals of socket-outlets and joint boxes (if any) connected into the ring and returning to the same way of the distribution board. Thus, each socket-outlet is fed from two directions which means that the effective area of conductor is doubled; also that a socket-outlet will not be disconnected if a break occurs on either side of it. Unless a domestic ring final subcircuit is in metallic conduit or is run in metal-sheathed cable, an earth-continuity conductor is also run in the form of a ring having both ends connected to earth at the distribution board (or its equivalent).

The minimum conductor size used throughout for a ring final subcircuit is 2·5 mm^2 if rubber- or p.v.c.-insulated, or 1·5 mm^2 if mineral-insulated. The protecting fuse must be rated at 30 A.

Socket-outlets used on domestic ring final subcircuits are shuttered type of 13 A rating to BS 1363 used in conjunction with cartridge fused plugs having flat pins. Two alternative fuse sizes are available: 3 A and 13 A.

In domestic premises a ring final subcircuit may serve an unlimited number of points, but must not serve an area of more than 100 m^2. Where two or more ring final subcircuits are installed, the socket-outlets and stationary appliances to be served should be reasonably distributed among the separate ring final subcircuits.

Nondomestic ring final subcircuits may use socket-outlets rated higher than 13 A and complying with BS 196. These have raised

72

socket keys to prevent insertion of nonfused plugs, together with socket keyways recessed at a specified position. The general requirement is that the number of such socket-outlets served must be limited so that in normal use the loading of the final subcircuit does not exceed the rating of the device protecting the final subcircuits.

Among advantages claimed for ring final subcircuits over radial final subcircuits are: an adequate number of outlets can be provided with economy of wiring, and individual flexible cords can be protected.

What is a spur?

A spur is a branch cable connected to a ring final subcircuit (Fig. 45). A subcircuit which includes spurs must be so designed that the total number of spurs does not exceed the total number of socket-outlets and stationary appliances connected directly in the ring.

For a domestic ring final subcircuit, spurs may be either fused or unfused, switched or unswitched (Fig. 46). Fused spurs must be

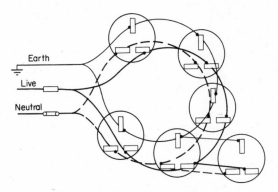

Fig. 45. Ring final subcircuit with spur

connected through the type of connecting box known as a fused spur box. The rating of the fuse in the spur box must not exceed that of the cable forming the spur and should never exceed 13 A in any event. Moreover, the total current demand of points served by a fused spur may not exceed 13 A.

Nonfused spurs may be connected to a domestic ring final sub-circuit at the terminals of socket-outlets, at joint boxes or at the

Fig. 46. Switched fused spur boxes
(Nettle Accessories Ltd.)

origin of the ring in the distribution board. They must have a current rating which is not less than that of the conductors forming the ring final subcircuit. Not more than two socket-outlets or one twin socket-outlet or one stationary appliance must be fed from a nonfused spur.

What is space heating?

This term denotes the heating of rooms, areas and people and things within them. It distinguishes this from other forms of heating, such as water heating, soil heating, etc. In general, the heat is transmitted by either convection or radiation. Convection relies on the creation of currents of warm air which rise and are replaced by cooler air (warm air being lighter than cold air). In radiation, which is the method by which the Earth receives heat from the Sun, objects are heated directly without relying on the heating of the intervening air.

What are convector heaters?

A common type of convector heater consists of an element contained within a casing having an inlet at the bottom for admission of cold air and an outlet at the top for the emission of warm air. It may rely entirely on natural air convection currents, or may be fan-assisted.

What are radiant convector fires?

They are heaters designed to combine direct heating by radiation with background heating by convection. Thus, they can be used to maintain a comfortable general room temperature while possessing the additional facility of supplying rapid localised heating if and when required.

What are tubular and skirting heaters?

Tubular heaters are circular or oval pipes of standard lengths containing electric elements. They can be installed singly or in banks of two or three, usually on a skirting.

Skirting heaters are also supplied in standard lengths for fixing to a skirting, but are usually designed in a similar manner to convector heaters with cold air inlet and warm air outlet.

What are panel heaters?

There are various types of electric panels, varying from the element embedded in heatproof insulating material (such as an asbestos composition) to the metal kind looking rather like a hot-water radiator but containing no liquid.

What are oil-filled electric radiators?

These have metal cases containing oil which is heated electrically. The oil is used merely to distribute the heat. The temperature is usually controllable.

What are fan heaters?

In these, a rapid distribution of heat is brought about by electric fans which blow air through the elements.

What are block storage heaters?

They are heaters comprising containers within which electric heating elements are surrounded by refractory material and insulation. They are designed so that electrical energy supplied during off-peak hours (mostly at night) can raise the temperature of the refractory material so that it can give out heat during the following day. In some types the output is controlled by fans.

What is the Electricaire system?

It is a system by which heat supplied to refractory material is given out in the form of ducted warm air to the various rooms from a central unit. A typical control unit is shown in Fig. 47.

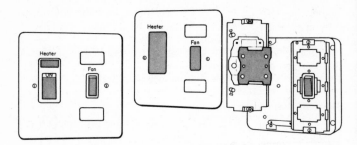

Fig. 47. Electricaire control units, 45A double-pole (J. A. Crabtree and Co. Ltd.)

What is floor warming?

This is a method of space heating in which grids of heating cables are installed in the floor. It is applicable to off-peak heating.

How are space heaters connected?

Portable electric fires, such as reflector fires and unit fan heaters, are usually connected to socket-outlets. Panel and skirting heaters may be fed from the fixed wiring via switched fused connection boxes.

Off-peak systems of space heating must be separately wired and controlled, as they are supplied with energy mainly at night and the electrical energy they consume is charged at a cheap rate. Both time switch and thermostatic control is incorporated. Thus, they are generally considered for wiring purposes as fixed appliances. If there is a special reason why they need to be connected via socket-outlets and plugs, these must be noninterchangeable with any other socket-outlets and plugs used in the installation.

What are 'industrial' methods of heating?

Methods of heating used in industrial processes include the following:

(*a*) Resistance heating, in which the material is heated either by an electric current being passed through it, or by being in contact with or irradiated from a device heated by a resistance element.

(*b*) Arc heating, where heat is produced as the result of an arc struck between an electrode and other metal.

(*c*) High-frequency induction, which involves heating a conducting material by the transformer effect. Heating is induced in the material by placing it within a coil through which a high-frequency electric current is passed.

(*d*) High-frequency dielectric, in which an insulating material is placed between two plates connected to a very-high-frequency electricity supply so that the material becomes heated by creation of rapid molecular disturbance within it.

What is an immersion element?

It is an electrical resistance type element used to heat liquids. Basically, it consists of high-resistance wire contained in, but insulated from, an outer sheath generally of metal.

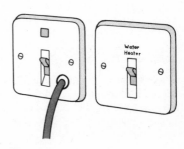

Fig. 48. Water heater switches
(Nettle Accessories Ltd.)

In a made-up electric water heater, the immersion element is fitted by the manufacturer, but when it is required to provide electric heating to an existing hot-water system (e.g. solid fuel back boiler), the system must be drained and the cylinder or other hot-water container drilled to accommodate an immersion element and thermostat. It is preferable to derive the electricity supply from a separate subcircuit with its own labelled switch (Fig. 48).

What is a nonpressure, or free-outlet, water heater?

It is the type of water heater often installed over sinks and wash-basins where relatively small quantities of hot water (e.g. 5 to 15 litres) are required from a single outlet. A typical heater of this type consists of a lagged metal container with a vertical electric immersion element. The water inlet is tap-controlled, so that hot water flows through an outlet pipe when the inlet tap is opened. The outlet must not be controlled, and arrangements are incorporated to prevent excessive rise in internal pressure (e.g. by a vent in the top of the container with a plug). Some patterns have an anti-drip device.

What is a pressure type of electric water heater?

It is the type of conventional hot-water storage system in which cold water in a tank at high level (e.g. in the loft) provides a head of water, the heating being carried out in a cistern or boiler fitted with an element and thermostat. The inlet to the tank is controlled by a float ball valve. The hot-water outlets at baths, sinks, basins, etc., are tap-controlled.

In some such systems, dual elements are fitted, so that either or both elements can be used. There is also a type specially adapted for installation underneath kitchen draining boards.

What are 'off-peak' electric water heaters?

These are generally multi-point heaters designed to take their heat up to 16 hours a day out of the 24. Basically, they are of two kinds, the pressure type and the cistern type.

It is important that there should be no appreciable mixing of incoming cold water with hot water. In one particular type of

78

pressure off-peak water heater, the container is deliberately made tall and slim, so that it entails a minimum area of contact between cold and hot water. In another type there are two cylinders, one on top of the other, connected by a short pipe. Each cylinder has a horizontal thermostatically controlled immersion element. The elements operate alternately. During off-peak times the top heating element is switched on first. Only when all of the water in the top cylinder is heated to the required temperature is the supply switched to the lower element. Thus, even when the lower cylinder is completely cold, there is fully heated water in the top cylinder.

With a cistern type off-peak water heater, hot water flows to the taps due to gravity, which avoids the necessity of displacing it by the admission of cold water. In one method, a feed cistern arranged above the storage container is controlled by an electromagnetic valve connected to the off-peak electricity supply, so that the valve is open only when the current is flowing.

Why is thermal lagging important?

It is important for almost all heating systems that there should be as little heat loss as possible. For an efficient hot-water supply system, it is essential that stored hot water should not lose an appreciable quantity of heat. This involves the provision of a lagging jacket around the storage vessel and effective thermal insulation around hot-water pipes. An indication of the effect of good lagging is given by considering that an unlagged hot-water cylinder will lose five times the heat that would be lost by one effectively lagged with glass fibre 250 mm thick.

What is an electrode boiler?

It is an efficient but rather expensive method of water heating in which heat is generated by the passage of alternating electric current between electrodes immersed in the water to be heated. The electrodes are vertically mounted cast iron with a neutral shield surrounding each electrode. It is essential for the outer case of the boiler to be effectively earthed and bonded to the sheath and armour of the service cable. Certain I.E.E. Regulations apply specifically to electrode boiler installations.

What is a heat pump?

A heat pump is an electrically operated device designed to extract heat from its surroundings and to deliver it elsewhere where required. For instance, it can be used to take heat from a larder and deliver it to a hot-water system. It is relatively expensive in first cost.

What is the load of a domestic cooker?

This differs enormously and may vary from 3 to well over 100 kW. Typical loadings of the different cooker parts are: oven 2·5 kW, grill 2·75 kW and four radiant hob rings at 2·75 or 1·5 kW each.

How are cookers connected?

A free-standing electric cooker is generally connected to the supply via a cooker control unit (Fig. 49) which consists of a switch of adequate capacity with visual on and off indication and sometimes a separately controlled socket-outlet. It is good practice to provide a separate cooker final subcircuit.

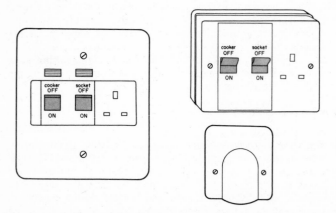

Fig. 49. Cooker control units (M.K. Electric Ltd.)

Diversity may be applied in accordance with the I.E.E. Regulations to determine the demand of fixed cooking appliances in domestic premises. For example, in the case of a cooker having a maximum load of 40 kW, the demand to be assumed in designing the subcircuit would be as follows:

the first 10 A at 100 per cent . . . . 10 A
30 per cent of the remaining 30 A . . . 9 A
assumed current demand 19 A

If a cooker control unit incorporates a socket-outlet for an electric kettle etc., another 5 A should be added to the current demand, making the total 24 A.

It is permissible in domestic premises for a final subcircuit having a rating of between 15 A and 30 A to supply two or more cooking appliances when these are installed in one room. For, apart from free-standing electric cookers, there are smaller cooking appliances such as table-top cookers, rotisseries, etc.

How does a self-cleaning cooker operate?

When the control switch is set to 'clean', the oven heats up to about 450°. At this temperature, cooking grease and splatter disintegrate, leaving a small amount of ash at the base of the oven. To prevent the oven from being opened while cleaning is in progress, there is a three-way safety interlock—a time lock, a thermal lock and a mechanical lock.

How does an infra-red cooker operate?

The element in this type of cooker is designed to produce the maximum of invisible infra-red radiation which, as we have already seen in dealing with radiant heating, can heat things directly without having to wait for the air to transmit the heat. This has the advantage of even cooking, so that the inner portion of food being roasted can be heated at the same time and to the same extent as the portions at the surface.

What is a fan oven?

As the name implies, it is one that includes a small fan which evenly circulates the hot air inside the oven and around the food; thus heating is quicker and more efficient.

What is a simmerstat?

It is a method of controlling the heat output of a boiling ring by automatic on and off switching of the element. The device includes a bimetallic switch operated by a small built-in heater. When the simmerstat is set at a low level, the element is on for short periods and off for long periods. When set at high level, the element is on for long periods and off for short periods. There are, of course, some intermediate settings.

What is a pan stat?

This is a thermostatic device which is fitted at the centre of a boiling ring to keep the temperature at the bottom of a pan at a preset constant temperature within limits.

What is an autotimer?

It is a clock control by means of which it is possible to preset the time for switching on and also the duration of cooking. Thus, a meal can be cooked automatically to be ready for serving at a given time.

What is a monotimer?

This is a short duration alarm system which is usually operational for periods up to one hour. For instance, it can be used to ring when eggs have been boiled for the desired time.

6

MOTORS

What is an electric motor?

An electric motor is a machine for converting electrical energy into mechanical energy (e.g. in the form of rotation). The movement is due to interaction between the magnetic field due to current in the rotor conductors and that due to current in the stator windings. For the purposes of this chapter, small motors driving domestic electric appliances are excluded, since these are normally connected to socket-outlets.

What are the basic requirements regarding electric motor subcircuits?

With certain exemptions, every electric motor should be provided with a method of: (*a*) isolation from the supply; (*b*) protection against excess current both in the motor and in the wiring; (*c*) stopping and starting; (*d*) preventing automatic restarting after a stoppage caused by a drop in voltage or failure of the supply.

Examples of exemptions are relaxation of the need to prevent automatic restarting after a stoppage in cases where motors are arranged to be started at irregular intervals (e.g. a pump motor automatically controlled by a float switch) or where failure to restart might be dangerous (e.g. a motor driving a fan removing poisonous fumes).

Every motor with a rating above 0·37 kW must be provided with control apparatus incorporating a suitable protective device against excess current in the motor or in cable between the device and the motor. The supply undertaking should be consulted regarding starting arrangements for motors requiring heavy starting currents.

83

Cables carrying the starting, accelerating and load current of a motor should be of rating at least equal to the rated full-load current of the motor.

Where a starter is provided affording protection against excess current, the rating of the motor final subcircuit protective device may be up to twice that of the cable between the device and the starter.

Which factors affect the choice of a motor?

In choosing a motor for any given purpose, some, or all, of the following factors should be taken into account: (a) the supply available; (b) the load conditions; (c) the speed(s) required; (d) the form of protection necessary.

How does the supply affect the choice?

The standard public electricity supply in the U.K. is alternating current at 50 hertz frequency. However, there are still exceptional situations where a direct current supply is available. This is particularly useful where appreciable speed control is required, as d.c. motors are readily speed-controlled.

With a.c. motors, a three-phase type is always more efficient than the corresponding single-phase motor. Thus, a single-phase motor, other than a very small one, would normally only be chosen when a three-phase supply is not available.

What is meant by load conditions?

Obviously, the greater the load the motor has to deal with, the greater must be its kilowatt rating. Apart from the size of the load, there is the question of whether the motor is run continuously or for restricted periods. Because of this, the British Standard specifies two classes of rating: (a) the full-time rating, relating to the load for which a motor may run for an unlimited period, and (b) short-time ratings, relating to the load at which a motor may be operated for a stated period and conditions at a specified ambient temperature.

How is the speed determined?

For a d.c. motor the speed depends on the type of motor. But in the case of an induction motor, the speed depends on the number of poles and on the supply frequency. For instance, an a.c. motor with two sets of poles operated from a 50 hertz supply would have a natural, or synchronous, speed of

$$\frac{50}{2} = 25 \text{ rev/s}$$

The actual speed would be slightly less than this by a factor known as the slip.

In general, for a given load, the lower the speed the larger and more expensive is the motor. Thus, when low speeds are necessary it is usual to install conventional speed motors with reduction gear.

What is the 'form of protection'?

Motor enclosures are designed to suit the particular conditions in which they are intended to be installed. Where only skilled personnel have access, unenclosed or open-type motors with only handrails for protection may be installed, but for most situations some form of protection is required. This affects the ventilation and cooling of the motor and, to some extent, the installation.

What is a 'screen-protected' motor?

It is a motor in which portions not enclosed by the normal frame are covered by mesh screens which permit the entry of air but protect personnel from contact with live or rotating parts and the machine from damage by falling objects.

What is a 'drip-proof' motor?

In addition to wire mesh screens, this type of motor has louvres or canopies to prevent ingress of dripping or falling liquids.

What is a 'pipe- or duct-ventilated' motor?

It is, of course, a motor designed for high-temperature situations and is thus provided with a ventilation system in which air is conveyed by pipes or ducts. There are three main kinds. In a *self-ventilated* motor, cool air is drawn in to take the place of rising warm air. In a *forced draught* motor, cool air is blown into the machine. In an *induced draught* motor, cool air is drawn through the machine.

What is a commutator motor?

It is a motor provided with a commutator and brushes which act as a reversing switch. When the motor is connected to a d.c. supply, the commutator, which comprises a number of copper segments to which the coils of the armature are connected, ensures that the armature continues to rotate in the same direction even though its conductors pass field poles of opposite polarity.

D.C. motors are classified in Fig. 50 in accordance with the manner in which the field coils are connected to the armature,

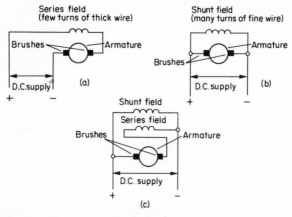

Fig. 50. D.C. motor connections: (a) *series,* (b) *shunt,* (c) *compound*

i.e. series, parallel (better known as shunt) or compound (a combination of series and shunt).

Where are series motors used?

They are used in electric traction and for some types of hoists. In general, they are chosen for varying speed drives requiring high starting torque at low starting speeds. They are usually direct-coupled to the load (i.e. not through belt drives), as a dangerous speed may be reached at very low load.

Where are shunt motors used?

These are commonly employed where relatively constant speed for all load conditions is required or where the speed will be constant for long periods.

Where are compound motors used?

There are two alternative methods of connecting the fields of compound motors, according to whether the series winding is arranged to assist (cumulative) or oppose (differential) the shunt winding. The cumulative type is used for drives involving high starting torque and fluctuating load. The differential type is used for special applications where it is necessary to have a tendency for the speed to rise as the load increases.

How is the speed of d.c. motors controlled?

The speed of a series motor can be varied by means of a variable resistor in series. A more effective method consists of connecting a variable 'diverter' resistor in parallel with the field, which has the effect of increasing the speed as the resistance is increased.

The speed of a shunt motor could be regulated by a variable resistor in series with the armature, but this results in considerable loss of power. A more convenient method is by a variable resistor in series with the field.

How is a single-phase a.c. motor made self-starting?

Three-phase motors of the *induction* type are self-starting as a result of the *rotating field* produced by the three phases reaching their maxima in sequence. But in a single-phase induction motor there is only one phase, and to cause the motor to start an artificial phase difference must be created.

One method, known as *split-phase*, consists of having two stator windings wound at right-angles to one another. One winding is of high resistance and the other is of low resistance. If a resistor, choke or capacitor is included in series with the high-resistance stator winding, the *starting winding*, the current in this will be further out of step with the main winding and the necessary condition for starting will be achieved. A centrifugal switch is used to disconnect the starting winding and capacitor or other element when the motor has run up to almost normal speed, as this winding is only light and would burn out if left in circuit.

What is a capacitor-start motor?

It is a type of split-phase motor in which a capacitor is used to introduce the required phase difference for starting (Fig. 51(a)).

What is a capacitor-start, capacitor-run motor?

This (Fig. 51(b)) is a single-phase motor in conjunction with which two capacitors are used, one purely for starting, and the other for permanent connection. Thus, not only is good starting torque provided, but since the starting winding remains in circuit while the motor is running, it improves the performance by causing it to operate rather like a two-phase motor. In addition, the permanent 'run' capacitor improves the power factor.

The capacitor-start, capacitor-run, circuit is mainly used for the larger sizes of single-phase motor.

What is a universal motor?

The name *universal* is given to the type of series field commutator motor used in small portable appliances, such as vacuum cleaners, drills, etc. It is suitable for use on alternating- as well as direct-current supplies because any reversal of the direction of current affects both field and armature windings, so that the motor continues to rotate in the same direction.

For use on a.c. supplies, the rotor and stator cores are laminated to reduce magnetic 'eddy current' losses. To improve the power factor and to minimise sparking at the brushes, a compensating coil at right-angles to the field coil may be provided.

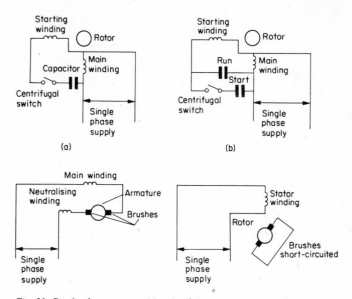

Fig. 51. Single-phase motors: (a) *split-phase capacitor-start,* (b) *capacitor-start capacitor-run,* (c) *series with neutralising winding,* (d) *repulsion motor*

What is a repulsion motor?

It is a commutator motor with the brushes short-circuited and arranged midway between the two limiting positions. The reaction between the rotor and stator field in this position results in a force of repulsion and consequently a torque which causes the rotor to revolve. This type of motor is suitable for fixed-speed drives requiring good starting torque.

What is a synchronous motor?

A.C. motors are either *synchronous*, in which the speed is dependent on the frequency of the supply, or *induction*, in which the speed may be affected by frequency, load or the position of brushes. Universal commutator motors do not come within either category.

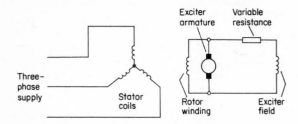

Fig. 52. Connections of synchronous motor

An essential feature of a synchronous motor (Fig. 52) is its constant speed. Small single-phase types are used in electric clocks. Three-phase types are sometimes designed to take leading current and are then suitable for improving the power factor of an installation.

What is a squirrel-cage motor?

It is an induction motor of simple construction. The rotor is made up of steel laminations and has slots in which are laid copper bars short-circuited by rings at each end (Fig. 53). The stator also consists of slotted steel laminations, but the slots contain windings.

A squirrel-cage motor is robust and reasonably cheap. Although basically of constant speed, it can be designed to run at different

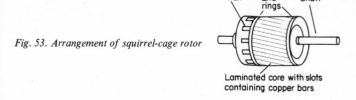

Fig. 53. Arrangement of squirrel-cage rotor

speeds by switching. A *double-cage* rotor provides a higher than normal starting torque.

What is a wound-rotor motor?

This is an induction motor (Fig. 54) which is provided with rotor windings connected to slip rings mounted on the shaft. Thus,

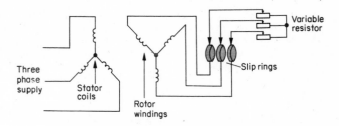

Fig. 54. Connections of wound-rotor induction motor

brushes arranged to contact the slip rings can connect external resistors in the circuit, so providing a relatively high starting torque from standstill. The value of external resistance is reduced as the motor builds up speed and at a pre-determined speed the brushes are lifted and the slip rings short-circuited. By suitable design of the resistors they can be used for speed variation in addition to starting.

Why is a starter necessary?

A starter is necessary for the larger motors (e.g. above about $\frac{1}{2}$ kW) to reduce the heavy current which would otherwise flow on first switching on. This heavy starting current could be a potential danger both to the consumer and to the supply undertaking concerned.

How does a direct-on starter operate?

In this method of starting (Fig. 55) the motor is directly connected to the supply without any device (e.g. resistor or inductor) for current limitation. The method is suitable for the smaller-sized motors only. A three-phase squirrel-cage motor, for example, takes about six times its normal full-load current on starting, so that in the larger sizes of motor it is imperative that this should be reduced. Direct-on starting is sometimes used for larger-sized a.c. motors in large factories where the heavy starting current can be absorbed.

What is undervolt, or no-volt, protection?

This is a form of protection designed to prevent automatic restarting of a motor in the event of an appreciable reduction in voltage or failure of the supply. In the case of a resistance starter, the contact arm is held by an electromagnet in the all-resistance-out position so that the arm returns to the off position by spring action on supply failure or on reduction in voltage. Unexpected automatic restarting by restoration of the supply after a failure could result in danger to personnel working on or with motor-driven machinery.

How is overcurrent protection provided in a starter?

In some starters there is a contact-maker operated by an electromagnet. The contacts are connected across the no-volt coil, so that when excessive current flows the no-volt coil is shorted out and the starter arm returns to its off position.

For a.c. motor starters, the overcurrent protective device may be either electromagnetic or thermal. The electromagnetic type consists basically of a soft iron plunger within a coil. Usually, movement of the plunger is restricted by oil so that it only operates on sustained

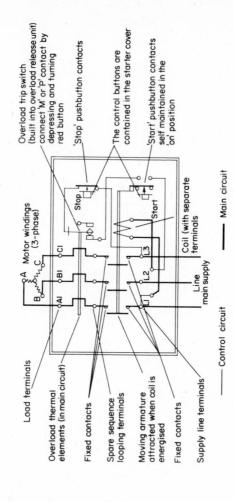

Load terminals

Overload thermal elements (in main circuit)

Fixed contacts

Spare sequence looping terminals

Moving armature attracted when coil is energised

Fixed contacts

Supply line terminals

Overload trip switch (built into overload release unit) connect 'M' or 'P' contact by depressing and turning red button

'Stop' pushbutton contacts

The control buttons are contained in the starter cover

'Start' pushbutton contacts self maintained in the 'on' position

Motor windings (3-phase)

A
B
C

C1
B1
A1

Stop

Start

Coil (with separate terminals)

L3
L2
L1

Line main supply

Stop

Start

— Control circuit   — Main circuit

Fig. 55. Basic start and stop pushbutton control for direct-on starter

overloads. The thermal type comprises a heating coil which operates a bimetallic strip at a preset value of excess current.

What is a star-delta starter?

It is one commonly employed for starting three-phase motors of the squirrel-cage type. Basically, the starter consists of a method of switching which connects the three stator windings in star for starting and in delta for normal running. The effect of connecting the windings in star is to reduce the applied voltage and current. This also results in a reduction in torque, so that the method can only be used when low starting torque is acceptable.

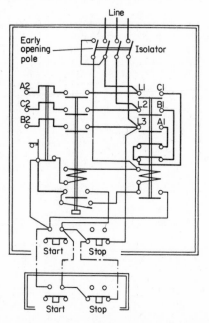

Fig. 56. Connections of automatic star-delta starter

94

The switching from star to delta connection may be carried out manually or automatically. In one type of automatic starter (Fig. 56) there are contactors with provision for overload protection and time delay.

What is the Wauchope method of starting?

It is a method (Fig. 57) used in conjunction with star-delta starting of avoiding switching transients during the changeover from star

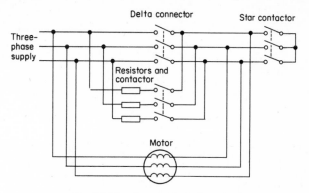

Fig. 57. Wauchope method of starting

to delta by the use of short-time rated resistors. Starting is achieved by a definite switching sequence designed to avoid a break in the supply current.

What is an autotransformer starter?

This is a two-stage method (Fig. 58) of starting three-phase squirrel-cage motors, but in this case the reduced voltage applied to the stator windings on starting is obtained via a transformer. Auto-transformer starters are more expensive than corresponding star-delta starters, but have the advantages of providing higher starting torque and of entailing only three connections in the motor terminal box instead of six.

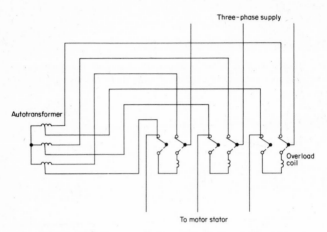

Fig. 58. Simplified diagram of connections of autotransformer starter

What is the Korndorfer method of starting?

It is an arrangement involving an automatic star-delta starter designed, like the Wauchope method, to minimise the current peak on changeover by avoiding a break in the circuit (Fig. 59). Three contactors are used. On closing the first starting contactor, the supply is applied to the stator windings via one section of an auto-transformer which thus acts as a choke. The motor then starts at low torque. The second contactor then closes, energises the auto-transformer, and applies a reduced voltage to the motor windings. When the motor has accelerated to a steady speed, the second starting contactor opens and the run contactor closes, applying full voltage to the windings. At this stage, the first contactor opens and isolates the autotransformer.

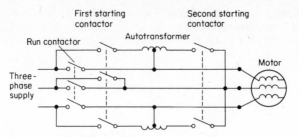

Fig. 59. Korndorfer method of starting

What is a primary resistance starter?

This is a method of starting three-phase three-terminal squirrel-cage motors under no-load or light load (Fig. 60). During the main portion of the accelerating period, fixed resistors are connected in series with the stator windings to reduce the applied voltage. As the rotor accelerates the starting current is reduced, thus giving less voltage drop in the resistors. This allows increased voltage to be applied to the stator. During the accelerating period the torque steadily increases. When the machine has attained the required speed, the resistors are short-circuited.

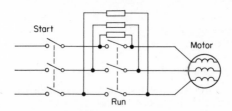

Fig. 60. Primary resistance method of starting

What is rotor-resistance starting?

It is the method used to start three-phase wound-rotor slip-ring motors, enabling the motor to start against full load (Fig. 61). By suitable design of the resistors in series with the rotor windings, the motor can have maximum torque on starting. In the hand-operated form, the starter includes a triple-pole switch with undervoltage

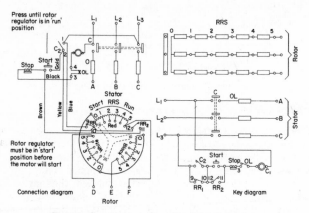

Fig. 61. Connections of rotor-resistance starter

and overcurrent releases for the stator circuit. The rotor circuit includes three suitably tapped resistors connected in star by a multiple tapping switch. The starter may incorporate an arrangement for short-circuiting the slip rings and lifting the brushes, so that the motor runs as a squirrel-cage motor.

How are motors reversed?

The direction of rotation of a d.c. motor (Fig. 62(a)) is reversed by reversing the current through either the armature or the field windings. Except for tiny d.c. motors with permanent magnet fields, reversing the supply leads to the motor will not alter the rotation, as it reverses the direction of the current through both armature and field. In order to avoid undue commutator sparking on reversal and possibly to maintain the torque, it is sometimes advisable to move the brushes.

To reverse a single-phase induction motor (Fig. 62(b)) the connections to one of the stator windings must be changed over.

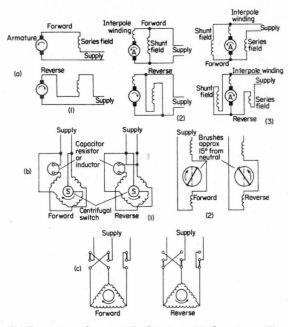

Fig. 62. *Connections for reversal of motors: (a) d.c. motors (1) series, (2) shunt, (3) compound; (b) single-phase motors (1) split-phase, (2) repulsion; (c) three-phase motors*

In the case of a universal motor, reversal of either the rotor or stator connections will alter the rotation. For a repulsion motor the direction of rotation is reversed by movement of the brushes round the commutator.

Three-phase motors are reversed (Fig. 62(c)) by changing over any two of the phase leads.

7

BELLS AND ALARM SYSTEMS

What is the principle of an electric bell?

The operation of an electric bell depends on an electromagnet (Fig. 63). The sound is due to the striking of a gong by a hammer caused to move by magnetic attraction. Usually, the electromagnet consists of two coils side by side wound on each leg of a common laminated iron core. A pivoted iron armature is arranged close to the

Fig. 63. Electromagnet

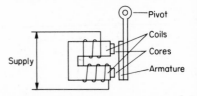

core ends so that when the electromagnet is energised it attracts the armature. The hammer is attached to the armature in such a way that it strikes the gong when attracted. There is a spring to restore the armature to its original position after attraction. The electromagnet coils are connected to the supply via a controlling push.

What is a single-stroke bell?

In a bell of this particular type, the gong would be struck once each time the push was pressed. Such a bell is therefore termed a *single-stroke bell*. It is useful for situations where a rudimentary form of signalling is required. For example, in public transport when the conductor uses one ring to tell the driver to stop and two rings to tell him to start.

What is a trembler bell?

A single-stroke bell would not be ideally suited to ordinary domestic or commercial purposes where one single ring might not be heard. For these situations, a *trembler bell*, which produces a continuous ringing tone as long as the push is pressed, is much more suitable (Fig. 64).

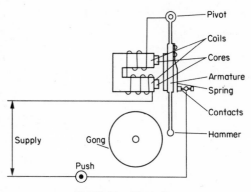

Fig. 64. Trembler bell

The trembler bell includes, in addition to the components of the single-stroke bell, a make-and-break contact in one lead to the coils. When the push is first pressed, the armature moves towards the electromagnet cores and the hammer strikes the gong. But in moving forward, the armature causes the contacts to separate, thus disconnecting the supply and, under the action of the spring, the armature is restored to its original position. With the armature restored, however, the contacts are remade, the electromagnet coils become energised again and the armature is attracted. This backwards and forwards 'trembler' movement of the armature continues as long as the push is pressed, producing repetitive strikes of the hammer against the gong.

101

What is a continuous-ringing bell?

For alarm systems it is usually insufficient to have a bell which rings only while a push is pressed. The continuous-ringing bell (Fig. 65) used for alarms is really a trembler bell with a spring-supported trigger and an additional set of contacts operated by an electromagnet.

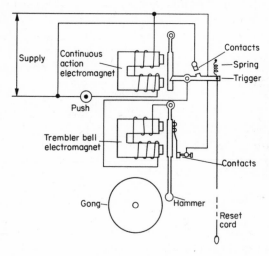

Fig. 65. Continous ringing bell

The armature of the continuous-ringing electromagnet contains a catch on which rests the pivoted trigger containing one of the two contacts. When the push is pressed the armature moves forward and the trigger drops. This causes the two contacts to make, and connects the bell directly to the supply (i.e. it shorts out the push or other contact device). The bell continues to ring as a trembler bell until a resetting cord is pulled to restore the trigger to its original position on the catch.

102

What are slow-beat bells?

They are alarm bells in which the striking mechanism is designed to operate slowly, thus giving a sound which can be distinguished from that of a trembler bell.

What are buzzers?

These are alternatives to a bell used for domestic and commercial applications, either where a sound less strident than that of a bell is required or where the sound must be easily distinguishable from that of a bell. A buzzer is constructed in a similar way to a trembler bell, but has no hammer or gong and relies on creation of sound by the vibration of the contact-breaker movement.

What are underdome bells?

In an underdome bell the movement is housed beneath the gong, making a more compact arrangement.

What is a motorised bell?

It is a bell in which a small electric motor replaces the electromagnet and armature. This type is much more expensive, both in first cost and in upkeep than a conventional electromagnet bell, but is used where a large volume of sound is necessary.

What are chimes?

Chimes are widely used in domestic situations due to their pleasing tones. Different types operate in different ways. For instance, double-note chimes are solenoid-operated with a spring-return striker designed to hit two tubes of different lengths. Another type will give three distinctive types of chime. Yet another is motor-driven and provides the eight-note tune of the Westminster chimes.

How are bell circuits wired?

Door bells for commerical and domestic use are usually operated at extra-low voltage (not exceeding 50 V between conductors and 30 V a.c. or 50 V d.c. between conductors and earth). In most cases the supply is derived from the 240 V public supply mains via a double-wound transformer. The wiring to the primary side of the transformer must be in accordance with the I.E.E. Regulations. The secondary wiring to bells and pushes is generally carried out in 0·75 mm² p.v.c.-insulated bell wire. This may be run on the surfaces of walls, ceilings, etc. Alternatively, it may be enclosed in steel conduit concealed in plaster, etc. It is important that extra-low voltage bell wiring should not be enclosed in the same conduit, trunking or duct with mains voltage wiring.

For certain applications, particularly alarms, bells operated at mains voltage are used. The whole of the wiring must then comply with the I.E.E. Regulations. Fire alarm circuits are often wired in mineral-insulated metal-sheathed cables, which are nonflammable, the supply being taken from a trickle-charged battery.

Which types of bell pushes are available?

Bell pushes are designed for various situations. Indoor pushes are usually in circular or rectangular moulded cases with contacts based on the 'springyness' of flat brass operated by direct pressure of the finger on pushbuttons. Pear pushes are used when they are to be attached to the end of a flexible cord. The weatherproof type of push is generally of the barrel pattern operated through a helical spring enclosed within a tube let into the wall or door frame. Front door pushes may have pushbuttons illuminated by low-consumption lamps.

What is the purpose of an indicator board?

A bell indicator board is generally installed when a bell is controlled from two or more pushes. The board then indicates which push has been pressed. Boards are made up into a number of 'ways', each way having a flag operated by an electromagnet having its coil in series with a push.

How does a pendulum indicator movement operate?

A pendulum indicator movement (Fig. 66) operates in a similar manner to a single-stroke bell, with a soft iron armature pivoted in front of an electromagnet. The armature is attached to the flag or other means of indication which is labelled or numbered and is arranged to swing in a window. When a pushbutton is pressed, the electromagnet of the relevant indicator way is energised and attracts

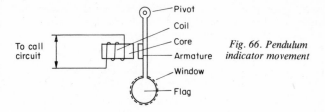

Fig. 66. Pendulum indicator movement

the armature and flag. When the pushbutton is released the armature is de-energised and the flag swings to and fro in the window.

Disadvantages of this type of indicator are that it cannot be used where it will be subject to vibration and that the source of the call is only indicated for a short period (i.e. until the flag has stopped swinging).

How does a mechanical replacement indicator movement operate?

The flag of a mechanical replacement bell indicator movement (Fig. 67) is displaced when the appropriate push is pressed and

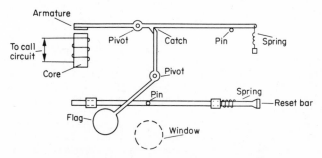

Fig. 67. Mechanical replacement indicator movement

remains in the displaced position until a mechanical reset bar is operated. Thus, there is no risk of the indication being lost by delay in looking at the indicator board.

On pressing a push, attraction of the armature towards the electromagnet releases a catch and causes the flag to indicate in the window. When the origin of the call has been noted, the reset bar is pushed and this restores the flag arm to its original position where it is held by the catch. There are several varieties of this type of movement.

How does an electrical replacement indicator movement operate?

There are two main types of electrical replacement indicator movement: (*a*) nonpolarised and (*b*) polarised. In the first type (Fig. 68), which is suitable for operation from either a d.c. or an a.c. supply, each indicator way has two electromagnets side by side, one in series with a call push and the other in series with a replacement push.

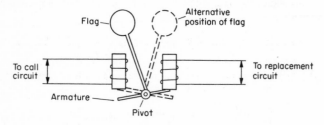

Fig. 68. Nonpolarised electrical replacement indicator movement

The armature and the arm holding the flag are pivoted between the two electromagnets. When the call push is pressed, the attraction of the armature causes the flag to move to one side. When the replacement push is pressed, movement of the armature causes the flag to move over to the other side.

In a polarised electrical replacement indicator movement, the call and replacement coils are wound on the same magnetised core. Thus, it is suitable for operation only from d.c. supply.

What is a luminous call system?

In some situations a silent call system (Fig. 69) is required. Audible indication by a bell is then replaced by visual indication such as

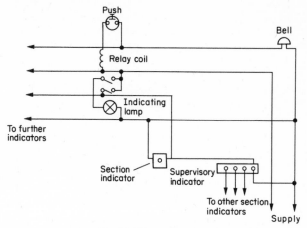

Fig. 69. Silent call system

lamps. For instance, in a hospital the system may incorporate luminous corridor indicators, a section indicator and supervisory indicators.

What are the requirements for a bell transformer?

Bell transformers (Fig. 70) for ordinary domestic and commercial call systems are usually designed for 230/240/250 volts tapped primary and 3/5/8 volts tapped secondary. They are double-wound, which means that their primary and secondary coils, although both wound on the same cores, are completely isolated from one another electrically.

Fuses may be provided on both the mains voltage and extra-low voltage sides. The transformer can be supplied from the mains via a socket-outlet. For the purposes of assessing total current demand of an installation, bell transformers take such minute current that they may be neglected.

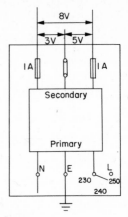

Fig. 70. Arrangement of bell transformer

What is an alarm system?

It is a system designed to give warning in the event of: (*a*) breaking-and-entering, or (*b*) the outbreak of fire. In a burglar alarm system, contacts are concealed at possible points of entry (e.g. doors and windows). In a fire alarm system, either special pushes can be provided or special automatic contacts can be installed.

What is a relay?

Basically, a relay (Fig. 71) consists of electromagnetically operated contacts. It is a device which enables one circuit (the *controlled*

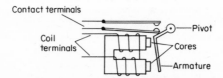

Fig. 71. Construction of typical relay

108

circuit) to be switched by means of another circuit (the *control* circuit). It has the advantage that the tiny current through the coil of an electromagnet can control a distant circuit operated by an independent supply.

How does an open-circuit alarm system operate?

This is illustrated in Fig. 72. It uses a type of relay in which the electromagnet when energised causes the relay contacts to come together. Instead of pushes, special contacts which are normally

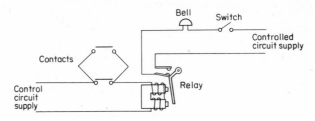

Fig. 72. Open-circuit alarm system

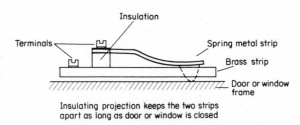

Fig. 73. Arrangement of open-circuit burglar alarm contact

open are fitted at doors, windows and other points of access. The contacts (Fig. 73) are included in the control coil circuit of the relay and are connected in parallel with one another so that closing of any one contact will cause operation of the bell. The disadvantage to the system is that it would become inoperative if the supply failed or if a burglar were to cut or disconnect the wiring.

How does a closed-circuit alarm system operate?

For a closed-circuit system (Fig. 74), the relay is designed so that the energised electromagnet keeps spring-loaded contacts apart.

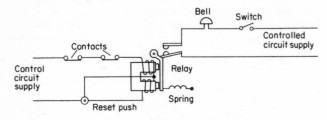

Fig. 74. Closed-circuit alarm system

Therefore, any break in the coil circuit should cause the contacts in the bell circuit to come together. The special contacts used are normally closed, so that opening the relevant doors or windows causes them to open. These contacts are connected in series.

What is a self-resetting burglar alarm?

If an ordinary closed-circuit relay were to be employed to control an alarm bell, the bell would ring continuously until switched off. There is a risk that, in switching off, the system may be left unintentionally inoperative. This is avoided by employing a self-setting arrangement incorporating a 'detent' (Fig. 75). Pressing a reset button then restores the relay armature to its on position. A test push is generally included in the reset circuit.

Provided that the contacts are closed, the relay armature is attracted to the electromagnet. But on the opening of a contact, the armature is released and falls down, thus completing the alarm bell circuit. When the call has been dealt with, the reset button on the relay is pressed, which causes the 'detent' arm to move into the vertical position, so holding the armature away from the contact in the bell circuit. The detent is locked in this position by the catch on the armature. However, the armature is not held at the full limit of travel, so that, when the contacts are all closed again, the armature

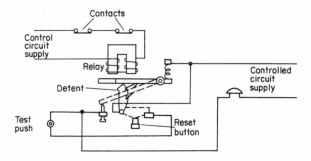

Fig. 75. Self-setting alarm circuit

lifts slightly and permits the detent to fall past the catch and return to the horizontal position.

For some premises, the door is open all day. In this case, a switch is not required in the coil circuit to the relay. The alarm is then automatically set on leaving the premises as the door is closed. Subsequent re-opening of the door will cause the alarm to operate.

What is the advantage of supplying an alarm bell from a battery?

The advantage is that the bell will then ring if the a.c. mains supply to the closed-circuit relax coil fails or if the voltage drops appreciably. Otherwise, the owner might be unaware that his system is in-operative.

How does a manual fire-alarm system operate?

The type of push used in a fire-alarm system is shown in Fig. 76. It is usually of the familiar glass-fronted pattern bearing the words: *In case of fire—break glass* in bold lettering. On the outbreak of fire, the glass must be broken to gain access to the push, and the

111

glass is often scored to facilitate this. For periodic testing, there is a separate test push available to avoid the necessity to break the glass.

In a simple open-circuit fire-alarm system, the alarm bell can be silenced by someone pressing the knob of a diversion relay. This

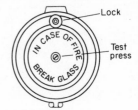

Fig. 76. Arrangement of break-glass fire alarm contact

locks electrically in the operated position until the contacts are re-opened and the switch restored in preparation for any subsequent alarm condition.

In a closed-circuit fire-alarm system, electric current flows in the circuit continuously. If an outbreak occurs, operation of the alarm contacts switches a high resistance into the circuit, and the alarm bell circuit is closed.

How do automatic fire-alarm systems operate?

There are various types of automatic fire detectors. Some of the electrical patterns depend on the unequal expansion of a bimetal strip resulting from excessive heat to operate the alarm contacts. The mercury type relies on the expansion of a column of mercury which is included in the circuit as a conductor.

A fusible-plug alarm incorporates a plug of low-melting point alloy. If the temperature rise is excessive, the alloy melts and releases a plunger which operates the alarm contacts.

Pneumatic alarm devices utilise the expansion of air in a sealed chamber. In the event of outbreak of fire, the expansion causes sufficient pressure on a diaphragm to force contacts to make.

Some fire-alarm systems, particularly those for the larger installations, include indicators to show the location of the outbreak.

What is a fire-call system?

This is a system sometimes used by firms which require their own fire service. It is designed to call fire-fighting personnel to an assembly point. It involves running a line to the residence of each fireman and installing a bell therein. At the headquarters there is a private switchboard at which all lines terminate and which is equipped with the necessary keys or plugs and jacks. Where all of the personnel live within the confines of, or near to, the premises concerned, an alarm siren may be installed instead.

How does an electronic alarm system operate?

One of the commonest types of electronic burglar alarm systems is that including a source of infra-red radiation focused onto a photo-electric device. The device usually incorporates a photoelectric cell in the grid circuit of a thermionic valve, on the anode side of which is connected a relay which controls the alarm bell. As long as the cell is irradiated, the alarm circuit is off. But any obstruction of the infra-red rays will cause the relay to operate and switch on the bell.

8

INSPECTION AND TESTING

What are the general requirements regarding inspection and testing?

Before being put into commission, every electrical installation and every major alteration to an existing installation should, on completion, be inspected and tested to verify as far as practicable that the requirements of the I.E.E. Regulations have been met. In the case of a major alteration, both the new work and that part of the existing installation relating to it should be inspected and tested.

In which sequence should the inspection and testing be carried out?

It is desirable that the sequence should be as follows:

    (*a*) Verification of polarity.
    (*b*) Tests of effectiveness of earthing.
    (*c*) Insulation resistance tests.
    (*d*) Test of ring-circuit continuity (if applicable).

What is involved in verifying polarity?

It should be ascertained that all fuses and single-pole control devices (e.g. switches, thermostats, etc.) are connected in the live conductor. Also that centre-contact bayonet and Edison-type screw lampholders in circuits having an earthed neutral conductor have their outer or screwed contacts connected to that conductor, and that wiring has been correctly connected to plugs and socket-outlets.

How is polarity verification carried out?

Provided that the installation concerned is connected to the supply, the check on fuses and single-pole control devices can be carried out using a test lamp of an approved type (e.g. a neon glow tester). In testing at lighting switch positions, lamps are removed and the tester, having a return connection to the neutral or earth-continuity conductor, is applied to the switch terminals (Fig. 77). In testing at

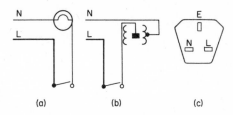

Fig. 77. Polarity: (a) *single-pole switch in live pole,* (b) *outer contact of centre-contact lampholder connected to neutral or earth,* (c) *connections of socket-outlet viewed from the front*

socket-outlets, a plug having a test lamp connected between the L and E terminals can be inserted in each socket-outlet in turn.

If the installation is not connected to the supply, either a tester incorporating an ohmmeter or a bell and battery set may be employed.

What is the purpose of a polarity test?

If a lighting switch is connected in the neutral conductor instead of in the live conductor, one terminal of the outlet will be permanently live whether the switch is on or off. If the live conductor is connected to the outer contact of a centre-contact or a screwed lampholder, the exposed metal of the lampholder will be live while the lamp is on, so that anyone touching it or any metal in contact with it may receive a severe shock. If the terminals of a socket-outlet are wrongly connected, metalwork of any appliance supplied from the socket-outlet could become live and consequently dangerous.

What is involved in testing the effectiveness of earthing?

A test of the earth-continuity conductor should be made (Fig. 78) to ensure that where earth-leakage protection relies on the operation of fuses or excess-current circuit-breakers, the impedance or resistance between the consumer's earthing terminal and the remote end

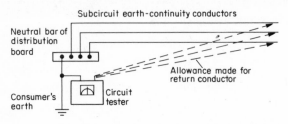

Fig. 78. Test of earth-continuity conductor

of every earth-continuity conductor does not exceed the maximum permitted by the I.E.E. Regulations in conjunction with the remainder of the fault path. The actual effectiveness of the earthing should be tested by means of an earth-loop impedance test in accordance with the Regulations.

How may a test of the earth-continuity conductor be made?

The I.E.E. Regulations recognise four alternative methods of testing earth continuity, as follows:

(a) A.C. test—This test is preferably made with alternating current at the frequency of the supply and of magnitude approaching 1·5 times the rating of the final subcircuit under test, except that currents need never exceed 25 A. The test should be made when the normal supply is disconnected from the final subcircuit being tested. One end of the earth-continuity conductor is connected to a cable of known resistance (this could be one of the final subcircuit cables).

The alternating voltage applied between the other cable ends should not exceed 30 V. For a satisfactory earth-continuity conductor, the ratio of voltage to current (after deduction of the resistance of the return conductor) should not exceed 1 ohm.

(b) Reduced a.c. test—As an alternative to *(a)* above, a hand tester capable of applying a lower value of alternating current at approximately supply frequency may be used for an a.c. installation. In this case a value not exceeding 0·5 ohm (after deducting the resistance of the return conductor) should be obtained where steel conduit forms part or whole of the earth-continuity conductor, or 1 ohm where the earth-continuity conductor is entirely of copper, copper-alloy or aluminium.

A hand-generator tester is often more convenient than one requiring connection to the supply, as it enables the live conductors while disconnected from the supply to be connected to the consumer's terminal and tested at various points between the live and earth-continuity conductor.

(c) D.C. test of a.c. installation—This can only be used if it has been ensured that no inductor is incorporated in the earth-continuity path. The acceptable values are as in *(b)* above.

(d) D.C. test of d.c. installation—For a d.c. installation the supply is conveniently obtained from a secondary battery and rheostat. The current should be approaching 1·5 times the rating of the subcircuit under test, subject to a maximum of 25 A. A hand tester applying less than full current may be used. The resistance of the earth-continuity conductor must not exceed 1 ohm.

What is involved in an earth-loop impedance test?

It is a test of the actual path which would be taken by fault current. The path includes the following parts, starting and ending at the point of fault: the earth-continuity conductor, the consumer's earthing terminal, earthing lead, metallic return path where available (this may consist of the metallic cable sheath or the continuous earth wire of an overhead line or, in the case of protective multiple earthing (p.m.e.) the neutral conductor or where no metallic return path is available the earth return path, the path through the earthed neutral point of the transformer winding, and the live conductor).

The test should be made with the cross-bonds required by the I.E.E. Regulations in place. The apparatus used for the test may show the loop-impedance in ohms. Alternatively, the testing equipment may merely give an indication of the maximum fuse rating or circuit-breaker setting, or may indicate whether a given rating of fuse or circuit-breaker setting will result in satisfactory operation under fault conditions.

Two alternative types of test are acceptable for making earth-loop impedance tests: (*a*) tests in neutral-earth loop, (*b*) tests in line-earth loop.

How is a neutral-earth loop impedance test carried out?

In this method, the test is made in the neutral-earth loop on the basis that the impedance of this path is the same as that of the line-earth loop (the actual path which would be taken by fault current).

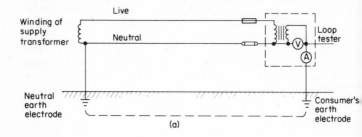

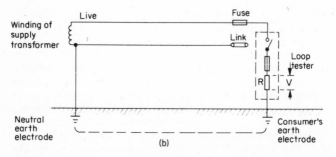

Fig. 79. Test of effectiveness of earthing: (a) neutral-earth loop test, (b) line-earth loop test

118

The apparatus is arranged to inject current into the neutral-earth loop, as shown in Fig. 79(a). The supply may be obtained via a transformer fed from the mains provided that no inductor is incorporated in the earth-continuity conductor, or from a d.c. source with its polarity rapidly and continuously reversed. The design of the instrument should be such that its indications are unaffected by neutral currents flowing in the system.

The I.E.E. Regulations require that when measurement on an a.c. system is made with less than 10 A and the earth-continuity conductor is wholly or mainly of steel conduit, the effective value should be taken as twice the measured value less the value at the consumer's earthing terminal. In all other cases, the effective value should be taken as the measured value less an appropriate deduction for impedances of the supply transformer, balancers, etc. This type of test must not be used on a system which is earthed by protective multiple earthing (p.m.e.).

How is a line-earth loop impedance test carried out?

The instrument used for this type of test is arranged to determine the current flowing when a known resistance (usually about 1 ohm) is connected between the live conductor and the consumer's earthing terminal, as in Fig. 79(b). The method has the advantage that it tests the actual path which would be taken by fault current. Also, it is suitable for use on systems earthed by p.m.e. In effect, an artificial earth condition is created and indication is based on the potential developed across the resistance. A typical instrument employs a resistance which causes a current of about 10 A to flow for 0·2 second.

Why is it necessary to test the effectiveness of earthing?

When an earth-leakage fault occurs with consequent flow of excess-current, it is important from a safety point of view that the current has an earth-loop path of low enough impedance to result in melting of the fuse or in opening of the circuit-breaker contacts. Thus, satisfactory operation of the protective device concerned depends on the effectiveness of earthing.

How is the effectiveness of an earth-leakage circuit-breaker tested?

The test consists of applying a voltage not exceeding 45 V, obtained via a transformer connected to the mains supply, across the neutral and earth terminals, or the neutral and frame terminals in the case of a voltage-operated earth-leakage circuit-breaker (Fig. 80(a)).

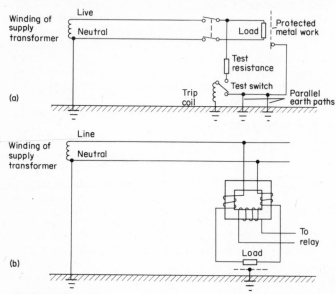

Fig. 80. Earth-leakage circuit-breakers: (a) *voltage-operated,* (b) *current-balance*

The transformer should preferably have a short-time rating of not less than 750 VA. The application of this voltage should cause the circuit-breaker to trip instantaneously. Where there is cross-bonding to other services, a direct earth connection may be introduced in parallel with the path through a voltage-operated earth-leakage circuit-breaker, but this should not prevent the circuit-breaker from providing adequate protection.

How is earth-electrode resistance tested?

Fig. 81 shows the method of connecting the test equipment. A steady value of alternating current is passed between the earth electrode X and an auxiliary earth electrode Y placed at such a distance from X that the resistance areas of the two electrodes do not overlap. Z is a second auxiliary earth electrode inserted midway between X and Y.

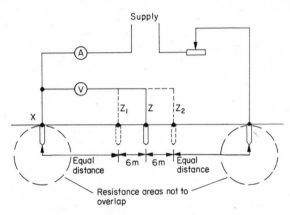

Fig. 81. Measurement of earth-electrode resistance (Appendix 4 of the I.E.E. Regulations)

The voltage drop between X and Z is measured. This voltage divided by the current flowing between X and Y is the resistance of the earth electrode.

As a check, two further readings are taken with Z moved 6 m nearer to X (Z_1) and then 6 m further from X (Z_2). If the three readings are substantially in agreement, the mean is taken as the true resistance. If they do not agree, the test is repeated with increased distance between X and Y.

The test can be made with current from the mains supply via a transformer using a high-resistance voltmeter (e.g. 200 ohms per volt), or with alternating current from an earth-testing instrument comprising a hand-driven generator, rectifier and direct-reading

121

ohmmeter. The earth electrode under test must be disconnected from all sources of supply other than that used for testing.

What is involved in insulation resistance testing?

An insulation resistance test measures the resistance of the radial path from a conductor through the insulation to surrounding metalwork, or through insulation between conductors. Contrary to conductor resistance, insulation resistance decreases with increase in length. Thus, an installation involving very long cable runs may have a naturally low insulation resistance, and for this reason the I.E.E. Regulations permit such an installation to be subdivided for the purpose of the test.

How are insulation resistance tests on an installation taken?

In making a test of the insulation resistance, direct current is passed through conductors and any leakage will be indicated by a relatively low reading. The testing voltage should be twice the normal r.m.s. voltage of the supply, subject to a maximum of 500 V for medium-voltage circuits. Direct current is used to preclude reactive effects. Tests are taken, to earth, with all poles or phases of the wiring electrically connected together (except for earth-concentric wiring), fuse links in place and switches closed (Fig. 82(a)), and between conductors (Fig. 82(b)). The minimum permissible insulation resistance is 1 megohm. If, in an installation containing a large number of outlets, the reading is less than 1 megohm, the installation can be divided into sections, each containing not less than 50 outlets, and the sections tested separately. Provided that each section test is not less than 1 megohm, the insulation resistance is considered satisfactory. Apparatus is disconnected while the test is made.

For a test to earth, all poles or phases of the wiring are connected to one terminal of the testing instrument, and the earth electrode is connected to the other terminal.

A test between conductors is made by connecting all conductors connected to any one pole or phase to one terminal of the instrument and conductors connected to other poles or phases of the supply in turn.

Where it would be impracticable to remove all lamps and to disconnect all current-using apparatus, it is permissible to test the insulation resistance with the lamps in place and apparatus connected, but with their local switches open.

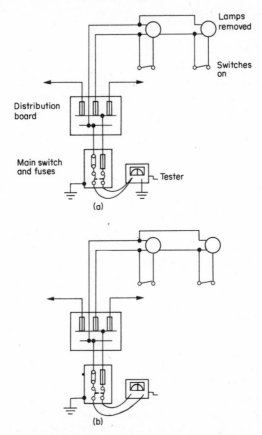

Fig. 82. Insulation test: (a) *to earth,* (b) *between conductors*

123

What is involved in insulation resistance tests on apparatus?

Fixed apparatus disconnected while insulation resistance tests are carried out on an installation must be tested separately. The test is taken between the case or framework of the apparatus and all live parts of each item of apparatus measured individually. The reading should be not less than 0·5 megohm for a satisfactory result.

Why are insulation resistance tests taken?

An insulation resistance test provides an indication of the effectiveness of the insulation in preventing leakage of current to adjacent metalwork and between conductors. Since the insulation resistance varies in accordance with such factors as ambient temperature and humidity, regular tests prove much more reliable information than an individual test, and it is recommended that a log should be kept of both insulation resistance and local conditions obtaining at the time of each test. In testing electric motors, it should be kept in mind that there are differences in the readings obtained while the machine is hot and those obtained when it is cold.

What is involved in testing ring circuit continuity?

This test is made to verify the continuity of all conductors, including the earth-continuity conductor, of every ring final subcircuit. In a new installation not connected to the supply, either a circuit-testing ohmmeter or a bell and battery set may be used. The ring subcircuit conductors are disconnected at the distribution board (or its equivalent) and both ends of each of the phase, neutral and earth-continuity conductors are connected in turn to the tester. In an installation connected to the supply mains, the test can be made by including a voltmeter or a test lamp in each subcircuit and earth-continuity conductor.

Why is ring subcircuit continuity tested?

The conductor size of ring subcircuits is adequate provided that each outlet is fed from two of them in parallel (i.e. the wiring is doubled up). But if there is a break or disconnection in the ring, one conductor will remain connected to the supply and could thus be overloaded. It is therefore important to ascertain that the sub-circuit conductors are looped into the terminals of socket-outlets and joint boxes (if any) connected into the ring and that they return to the same way of the distribution board.

What is a Completion Certificate?

It is a certificate which after inspection and testing is given to the person ordering work involving a completed installation or a major alteration to an existing installation. It is normally given by the contractor or other responsible person, but may in some cases be given by an authorised person. The certificate, as prescribed in the I.E.E. Regulations, certifies that the electrical installation has been inspected and tested in accordance with the requirements of the Regulations and complies (except for any departures listed therein) with the edition of the Regulations current at the date of contract for the work. The Completion Certificate does not cover portable

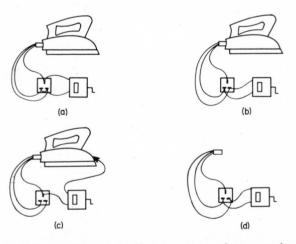

Fig. 83. Testing an appliance: (a) *insulation resistance between conductors and frame,* (b) *continuity of circuit conductors,* (c) *continuity of earth-continuity conductor,* (d) *insulation resistance of flexible cord*

appliances or apparatus connected to socket-outlets for which an Inspection Certificate may be obtained.

What is an Inspection Certificate?

It is a certificate which should always accompany, and be attached to, the Completion Certificate. It lists items inspected or tested with comments and departures (if any) from the I.E.E. Regulations.

How is a portable electrical appliance tested?

Tests on a portable electrical appliance include: (*a*) insulation resistance between conductors and frame, or case, (*b*) continuity of conductors, (*c*) continuity of earth-continuity conductor, (*d*) insulation resistance of flexible cord. The connections using a megohm/ohm testing instrument for these tests are illustrated in Fig. 83.

INDEX

127